BEN PHILLIPS

SORRY BRO!

Published by Blink Publishing
3.25, The Plaza,
535 Kings Road,
Chelsea Harbour,
London, SW10 0SZ

www.blinkpublishing.co.uk

facebook.com/blinkpublishing
twitter.com/blinkpublishing

HB – 978-1-911274-04-9
Signed edition – 978-1-911274-74-2
TPB – 978-1-911274-46-9
Ebook – 978-1-911274-05-6

A CIP catalogue of this book is available from the British Library.

Designed by Perfect Bound Ltd
Printed and bound in Italy

Pictures: all © 2016 Ben Phillips Global Limited except:
Alamy: Mediablitzimages 57. **Dan Rouse**: cover, 10, 30, 47, 52, 72, 76, 92, 99, 156, 160, 176, 178, 180, 202, 218-219.**Clive Sherlock**: 22, 42, 45-46, 62, 82, 94, 104-107, 114, 126-129, 136, 146, 168, 174-175, 191, 192, 196, 200, 209, 220-221. **Dreamstime**: Anthonycz 12; Csuzda 18, 19, 32-37; Subhan Baghirov 38, 48. **iStock**: Amanda Rohde 66, 67; hillwoman2 91. **Shutterstock**: 1000 Words 198; 123dartist 142-143, 145, 150; Adchariya Sudwiset 209; Africa Studio 91; Aksenova Natalya 187, 211; AlenKadr 117; Alex Brylov 68; Alexander Bark 38; Alexey Boldin 149; Alexey Romanowsky 152; AmazeinDesign 55-57; Andrey_Kuzmin 79-80; BackgroundStore 194; Bildagentur Zoonar GmbH 78; BonD80 142; Brent Walker 74; Brian C. Weed 18; c12 165; Claudio Divizia 96; CURAphotography 164; David Hughes 60, 110; Dragan Milovanovic 222; efiplus 125; Eric Isselee 59; Evgeniia Speshneva 60-61; exopixel 200; Freedom_Studio 198; godrick 216; GooDween123 60; gourmetphotography 33; gualtiero boffi 97; Hein Nouwens 87; Iakov Filimonov 40, 128-129, 214; ibreakstock 191, 206; Igor Kovalchuk 159; Igor Link 145; ilolab 20; In Green 205; iQoncept 172; irin-k 55; Ivan Baranov 86; Jazmin Gonzalez 116; Jes2u.photo 170; Joe Gough 91; Kagai19927 163, 180; kao 194; Katya_Branch 40-41, 86-87; kerenby 152; Kompaniets Taras 151; Kotin 100, 102; Ksenija Toyechkina 70; Leena Robinson 54; Lerche&Johnson 60; lynea 86; M. Unal Ozmen 79, 150; majeczka 58; Marcio Jose Bastos Silva 166; Marish 118, 120, 122, 139-141, 184, 214, 215; Mark Carrel 187; Marzolino 85; Mega Pixel 206; Michael C. Gray 75; MichaelJayBerlin 217; Mikhail Melnikov 24; mikser45 152-153; Morphart Creation 84-85; MyImages – Micha 188; Naatali 81; Natykach Nataliia 128, 215; OHishiapply 27; Ollyy 118-120, 139-141, 183-184; optimarc 55-57; Owain Davies 35; percom 36; phokin 122; photogal 210; Picsfive 100, 102-103, 122-123, 125, 164-165, 167, 209-211; Piyaset 180; Piyato 18-19; pjhpix 112; Ralf Beier 117; revers 151; Richard Griffin 91; Rudmer Zwerver 34; s_bukley 196; siloto 186, 188, 191; SpeedKingz 24- 25; Stefano Cavoretto 188; StudioSmart 150; Sutichak 66; Timmary 123; tobkatrina 58-61, 79-81; topseller 211; Twocoms 130; Valentin Agapov 74, 95; Volodymyr Krasyuk 167; welcomia 185; worker 20-21, 38-41; yukibockle 90.

1 3 5 7 9 10 8 6 4 2

Papers used by Blink Publishing are natural, recyclable products made from wood grown in sustainable forests. The manufacturing processes conform to the environmental regulations of the country of origin.

Blink Publishing is an imprint of the Bonnier Publishing Group
www.bonnierpublishing.co.uk

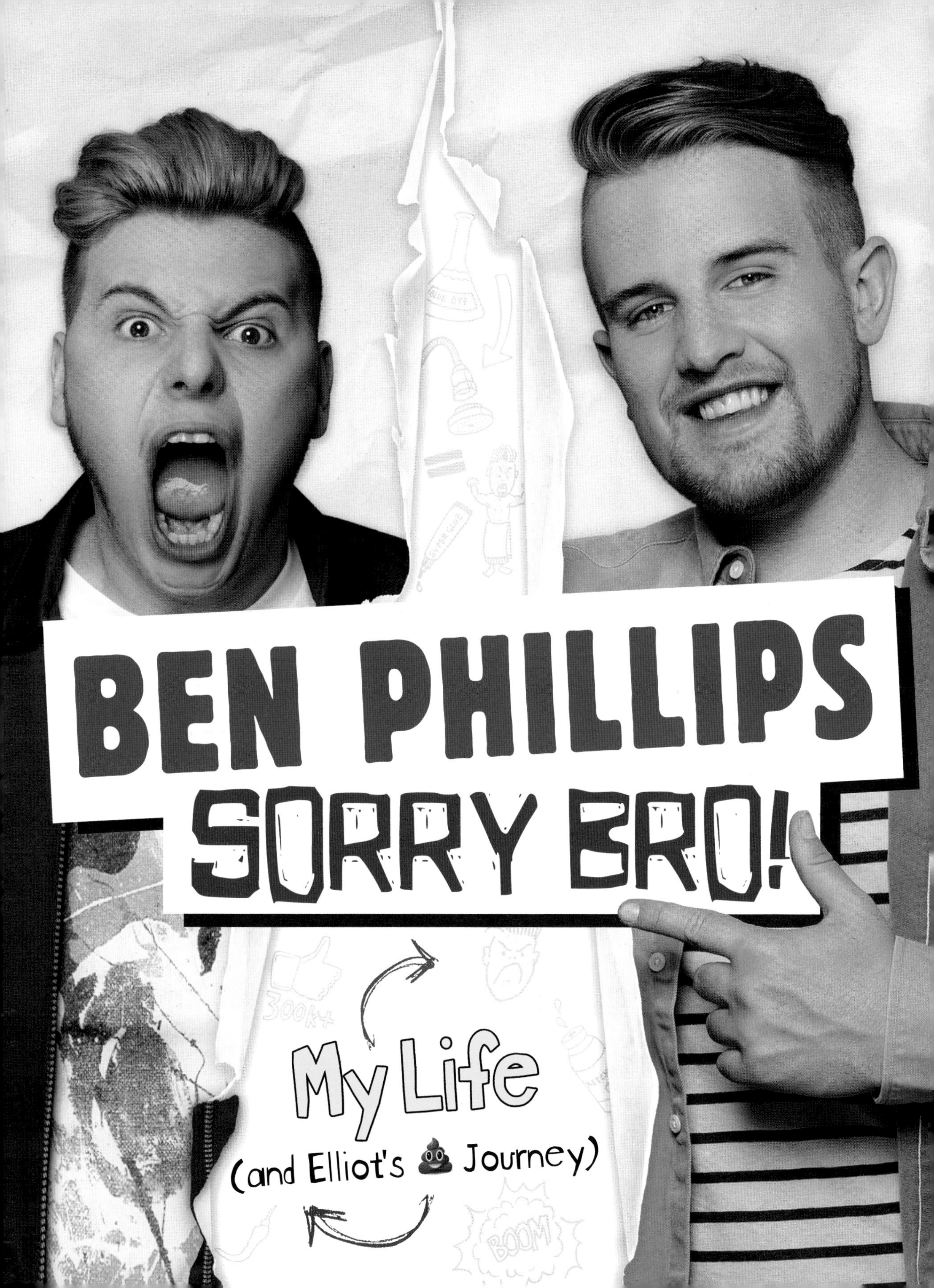

BEN PHILLIPS
SORRY BRO!
My Life
(and Elliot's 💩 Journey)

CONTENTS

DISCLAIMER
A WORD FROM BEN

This book is meant to entertain and make you laugh. A lot. The pranks you'll find in these pages are performed by professional donks and should NOT be copied under any circumstances. They're childish, often dangerous, and not funny unless your name is Elliot.

Despite how it looks, every effort has been taken to make sure my brother comes to no harm. His dignity might be beyond saving, and apart from some close shaves around the nipples, his personal safety is important to me. So, I'm asking you guys right now not to follow in my footsteps and attempt to copy anything you read in this book unless your name is Elliot Giles.

Collect All the Signatures

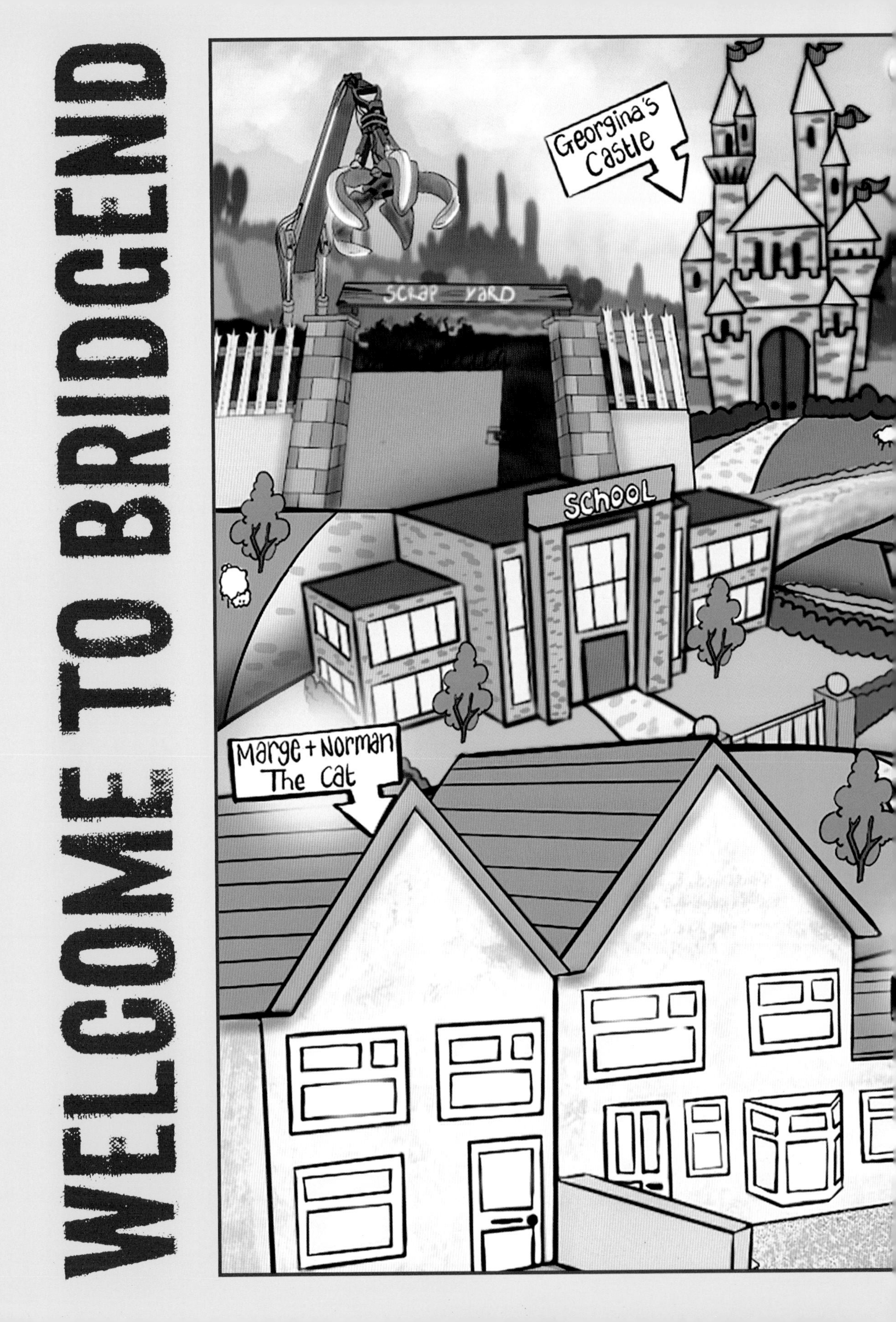
WELCOME TO BRIDGEND
Georgina's Castle
SCRAP YARD
SCHOOL
Marge + Norman The Cat

HERE BE DRAGONS
Farmer Chris
Nana's HOUSE
THAT WAY
THIS WAY

Chapter 1

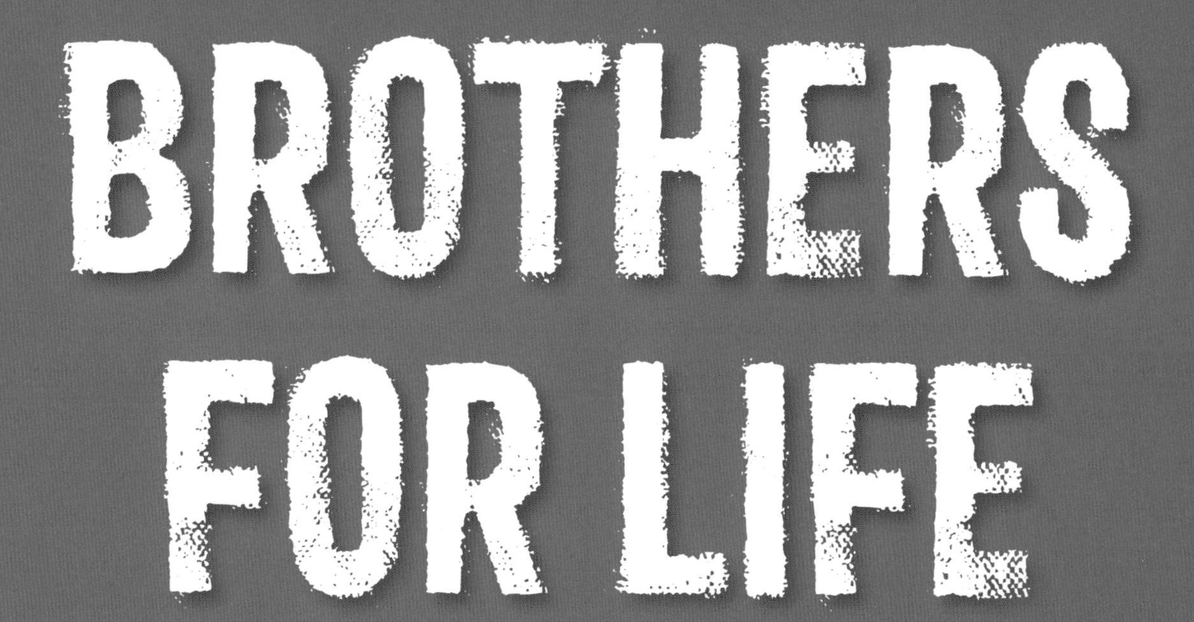

0

In the beginning, there was Ben. Yeah, that's me in the picture. Cute, huh? With a winning smile and rosy cheeks, I was everything my parents could've wished for. At least that's how I had it planned out as I learned to walk, talk and grasp a crayon in my little hand ...

Loved laughing back then too

I'm so cute

Now, I grew up in Bridgend. Anyone who's been to this pocket of Wales will know that the sun doesn't shine too much, but it seemed sunny to me as a kid. We were the perfect family, so I thought. I brought light and joy into my parents' world, and they rewarded me with cookies and sweets.

My first motor

Then something happened out of the blue. It was a chance event and we could've walked on by. Instead, my parents did what they believed to be the right thing, and it changed my life forever ...

One Sunday after lunch, Mum and Dad decided we should stretch our legs in a forest out of town. I was skipping ahead, kicking up leaves and singing to myself, when a rustling sound stopped me in my tracks. I spun around, spotted a creature no bigger than a badger scampering into the undergrowth, and gasped!

'Hello?' I squeaked. 'Who's there?'

A moment later, I spotted a pair of eyes blinking at me from a bush. They were big and pinched in fear at the corners, much like mine at that moment. I won't lie. Just then I was completely shocked. My parents were some distance behind me. Mum was admiring flowers and Dad was just, well ... standing there, probably wishing he was at home watching the rugby. Anyway, the fact was they weren't paying attention to me, which meant I was all on my own in the presence of an unseen *beast*.

'C . . . come out,' I piped up bravely. 'Whatever you are.'

The bush? It began to tremble and shake. I tried very hard not to do the same thing. I may even have done a little wee of fear in my shorts, but hey! I didn't turn on my heel and run. Looking back, I suppose that would've been the smart move. It might well have avoided years of grief, misery and pain (not for me, of course). Instead, my parents finally picked up on my cries and came running.

'What is it?' asked Dad as Mum scooped me into her arms.

Even though those strange eyes had withdrawn into the gloom, I raised my arm and pointed accusingly. In response, the quivering bush appeared to do a little wee of fear on the forest floor.

'Stay back,' Dad warned. 'It might be dangerous.'

Bowl cut

'I want go home!' I wailed, clinging to Mum as my dad collected a stick from the floor and approached the bush. 'Make it go away!'

He prodded the bush, and then stepped back smartly when it whimpered.

'Kill it!' I screeched at him. 'SET FIRE TO IT OR SOMETHING!'

'He'll do no such thing.' Mum set me down so smartly that I gasped in surprise. Watching her approach the bush, making gentle cooing sounds, I had never felt so abandoned in my life.

'Come back!' I pleaded, stretching out my hands. 'You can't just leave me like this!!'

Mum glanced back at me, irritated at my whining, and then crouched down to peer through the leaves.

'Don't be frightened, little thing,' she whispered. 'We won't harm you.'

'Yes, we will!' I insisted, and armed myself with the biggest stick that I could find.

Love cereal

'Son, that's enough!' Dad sounded cross all of a sudden, which left me no choice but to drop my weapon and step back. Meanwhile, Mum reached into the bush with both hands.

'Just be careful,' Dad cautioned. 'In case it bites.'

I admit a little part of me hoped that it might. Had a pair of jaws opened up behind the leaves then maybe Dad would've taken my advice and sent the mysterious beastie scurrying on its way. He might even have stamped on it, but that was wishful thinking.

Instead, still promising that we meant no harm, Mum grasped hold of the hidden creature with both hands and drew it out into the sunshine.

Me having a great time

'Put it down!' I yelled, and I didn't just mean on the forest floor. 'Daddy, I'm frightened!'

'Not as much as this tiny angel,' whispered Mum. She supported her find in both hands, raising it high into the light so that we could all see. Now, you have to remember I was only little. As far as I was concerned, that scrawny, hairy creature with boggle eyes and mud-encrusted face, fingernails and feet was a wolf cub, and I didn't like it one bit. Only later, when I was old enough to understand, did I learn that we had stumbled upon a wild boy – a child that had been lost or even abandoned in the forest and raised by kindly creatures. Dad said it was probably foxes. Looking back, I reckon earthworms played a role in his care. Earthworms and dung beetles.

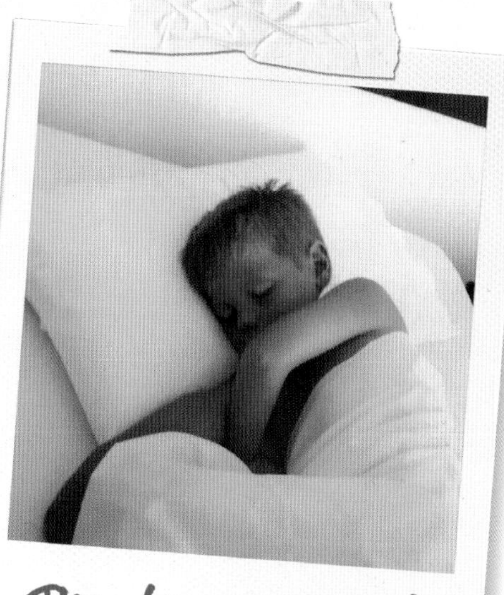

THAT'S MY BED!!

So, having set out for a nice walk that afternoon, we came back with a new addition to our family. Can you believe it? My parents could not do enough to make this mute scrap of a human being feel at home. Honestly, they reacted as if their discovery was destiny. I felt sick to my stomach, and no amount of tantrums would make them realise they were making a huge mistake. Through my eyes, it felt more like a soap-dodging ratboy with fleas had moved into my life.

That week, I was not a happy little boy.

Ben's direy

Mundai

WULF has Kam to live wiv us. Wulf is hairy. He maed the barth durty. I thinck mUmmie shavved him to mek him look nise. He dusnt' look nise. He is horride and just blinkes and blinkes lik he is stuypd! mUmmie gave him summ of MI cloths to ware. So HE IS A THEEF NOW TOO!!!! I hayte him!!!!

Chuwsdai

MUmmie an Daddee haf call him Eleyert. This maks me larf. It sownds lik ELERFERNT, an we all nowe thay r big an stewpid. Hahahaha! Elerfernt!!! Hopfelly he will be sent bak to the wuds soone an I can ferget abowt him.

~~Wen Wedz Weners~~ The nex dai

Now mUmmie an Daddee sai thay r adupting him. Wot? I dono want this meens. I hope it meens they wil droun him in a buket. Hey, Elyert – haf a happy adupton! Nise nowing u budee!!!

Thrusdai

Elyert is stil heer!! He Keeps lukin at me lik he dusnt truste me. Its so annoing!! I haf told him he Kan go bak to the wuds any tyme. He jus blinkes at me. They say a puppee is not jus faw Krismas, but if Elyert was a dog he wud be packin his bages on BoKsing Daye.

Daddee sai 'be nise to ur bruther, Ben.' But Elyert is nut my bruther!!! He is a shavved wulf who needs adupting fast. I even fownd a buket in the shed so they cud do it. mUmee jus sed 'wot a Kind bruther u ar, Ben. Wy don't u both playe in the gurdun? So I tuk Elyert into the gurdun an filled the buket wiv warter frum the hoze. But his hed wus to big to fit in the buket and I cud not adupt him. But I did mayk him cry. Wich was fun!

Frdyai

Elyert dusent sai much. He jus grunts an makes a fase lik he is duing a pooe. mUmmee sai we haf tu teech him tu tork. An so tuday I teeched Elyert his ferst wurd. All I dun wus put marbels on the flaw owtside his room.

Then I invertated him to playe.

'Elyert!' I sed. 'Wood yu lik to pley a funnee game?' I hurd him grunt. The he erpened the door, stepped out and went flyering. It was grate! Wen Elyert stopped being lazy an unconshus, he shooke his hed, all red in the fayce lik a turmarter an yeled 'Ben..Ben! BEN! **BEN!**'

mUmmee was so happee to heare him speekeng! She cam runinng up the sters, told Elyert to Karm down an stop chasying me. Then she hugged us berth so tite we cud berely breeth.

'I can cee u boys will git alung jus fine,' she sed. 'I hope wun day yu wil realyse how luKy u ar to haf eech uther.'

ELLIOT'S SECRET JOURNAL:

A CRY FOR HELP

Two of my kidnappers are all right, y'know? In a weird way, they treat me as one of their own. The third? Well, they keep on saying he's my brother but all I know is he's trouble.

Normally, when you're snatched in broad daylight, you expect to be bundled into the back of a van at gunpoint. In my case, the gang leaders have shown nothing but kindness and respect. After washing away my woodland pong (which I thought was kinda funky in a good way), and shearing my winter coat with clippers, the pair sorted me out with a freshly made bed, three decent meals a day and access to a hypnotising machine called a TV. I'll be up front with you here. Life here is better than the burrow that I used to call home. No doubt my furry family fled to safer fields just as soon as the kidnappers seized me, and I hope things work out for them, I do. Above all, I'll miss the wise owls, who taught me everything from astronomy to woodland philosophy, and though I'm new to the language of my captors, I'm learning fast.

In fact, this new life would suit me just fine were it not for the boy called Ben. There's a glint in his eye that causes what hair is left on the back of my neck to rise, while his laugh is enough to send a chill right through me. In short, both are signs that I'm set to suffer at his hands.

I can only hope that it's just a phase. With luck, he'll grow tired of poking me with sticks and stuff. In a short while from now I might even look back upon this time and laugh.

Ben Phillips @benphillipsuk

And d'you know what? In a weird way that's exactly how it all turned out. I could cut to the chase and put you in the picture, but that would mean missing out on all the pranks at Elliot's expense. So, let's go on with the story that brought us to the attention of the whole wide world ...

It's not that we don't like you, bro...

INSIDE THE BRAINS OF **BEN** AND **ELLIOT**

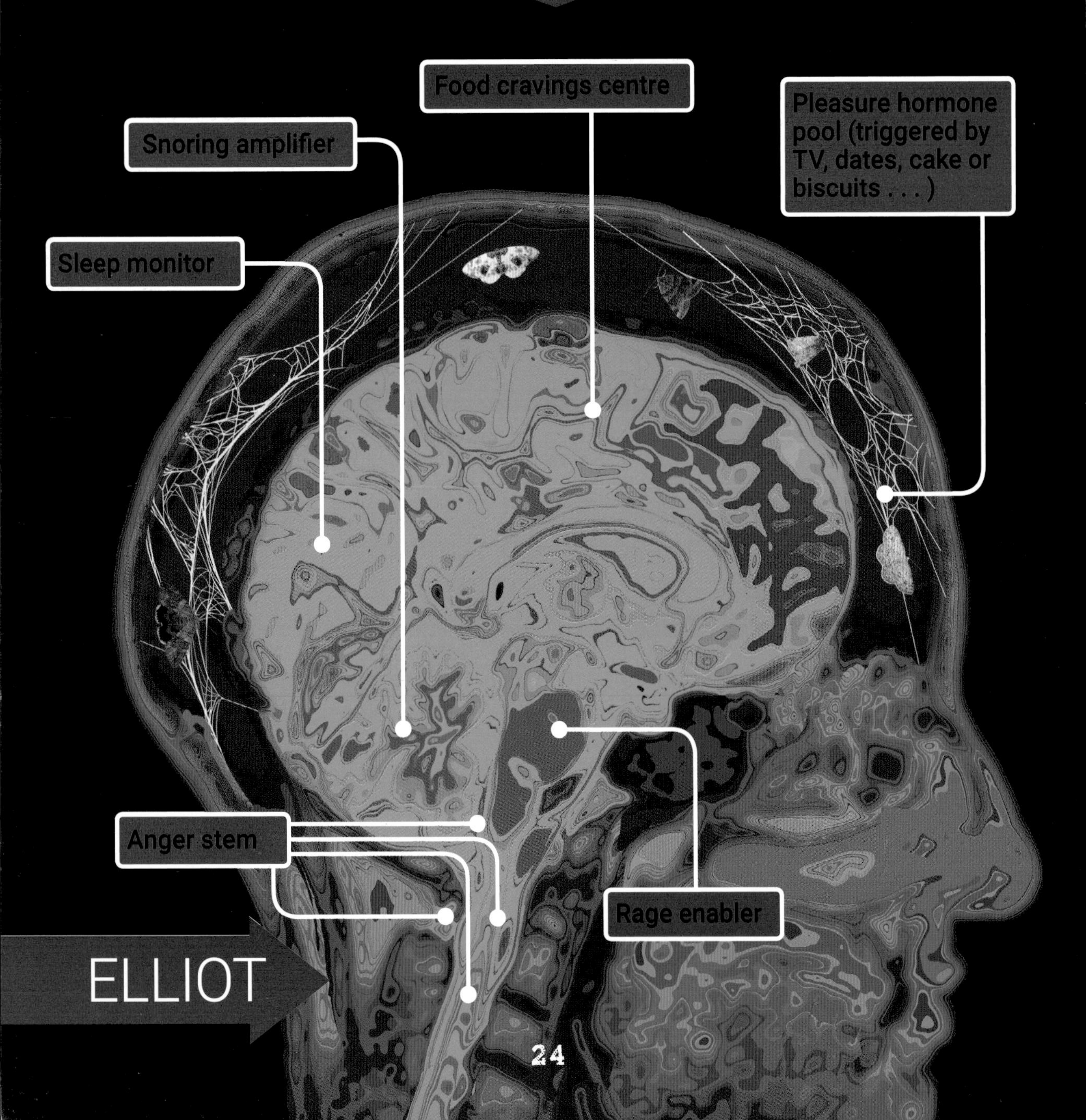

We might be brothers, but Elliot and I think very differently. So the last time we visited A&E, to get the plastic unicorn horn unstuck from his forehead, we nipped into the scanning department to get our heads examined. Here's what they found.

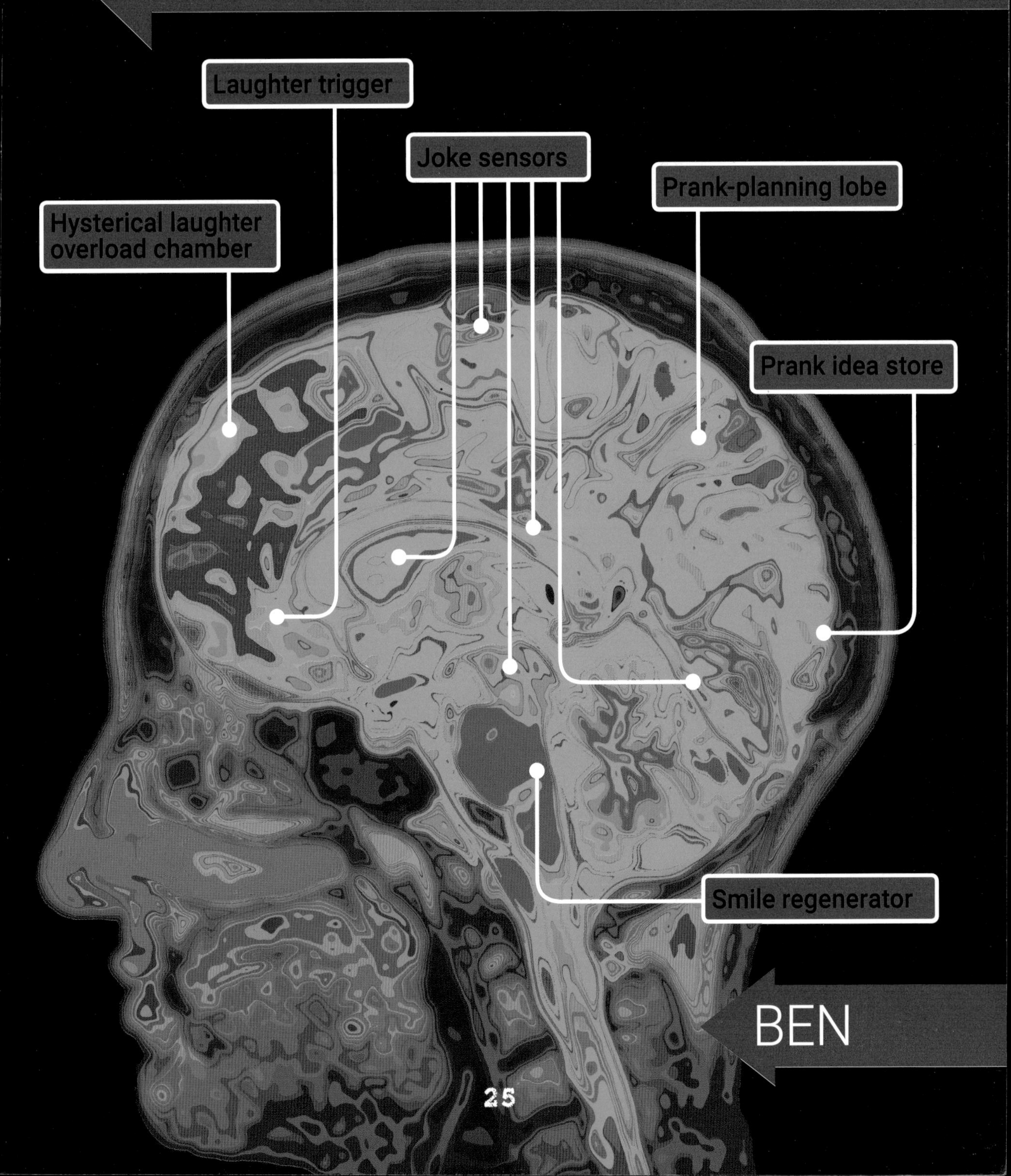

A few moments of happiness

amid the torture

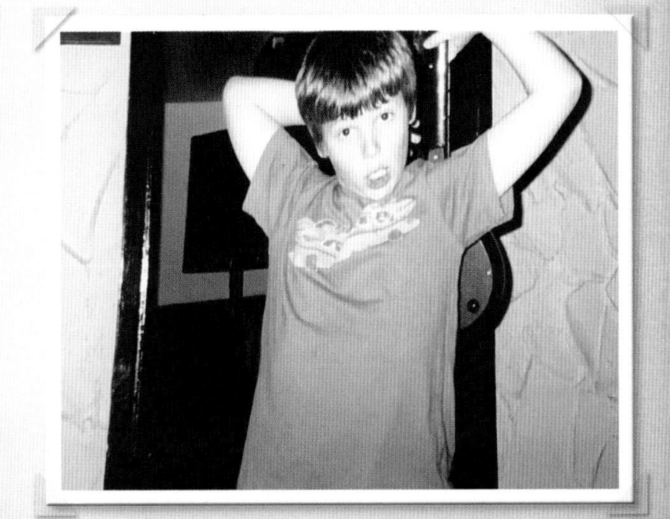

PERFECT YOUR PRANKING TECHNIQUE

After a lifetime of winding up Elliot, here are my top prank tips so everyone sees the funny side (eventually).

1 PICK YOUR PRANK VICTIM

Elliot might not laugh out loud when I stick him to the car seat, paint his face with permanent marker or glue stinging nettles to his nipples, but deep down I know he's one tough cookie. So before you spring your surprise, make sure your target can take a joke.

2 BE CREATIVE

If someone can see what you have in store, the joke's on you. It's all about understanding what makes your target tick and then using that knowledge to your advantage. Take Elliot, for example. My bro is always looking for love, and so I like to pounce just as he's stepping out on a date. When he grabs one last glance in the mirror, and finds the pox spots I've dotted all over his face with a permanent red marker, it'll be the last thing he needs at such an important moment – and that's where your efforts shine. So use your imagination and aim to pull off your prank as best as you can. They'll appreciate the dedication. Honestly!

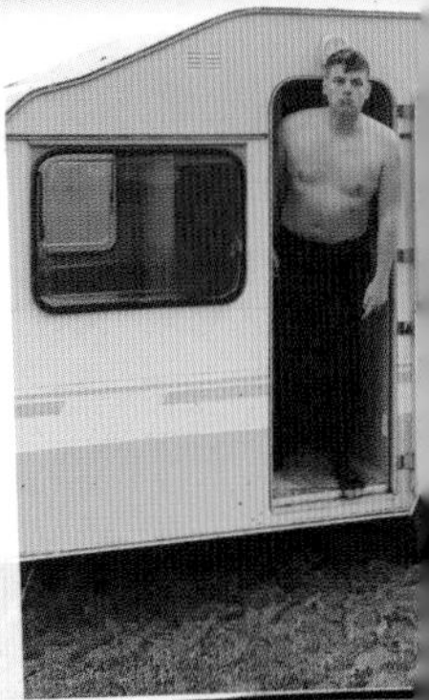

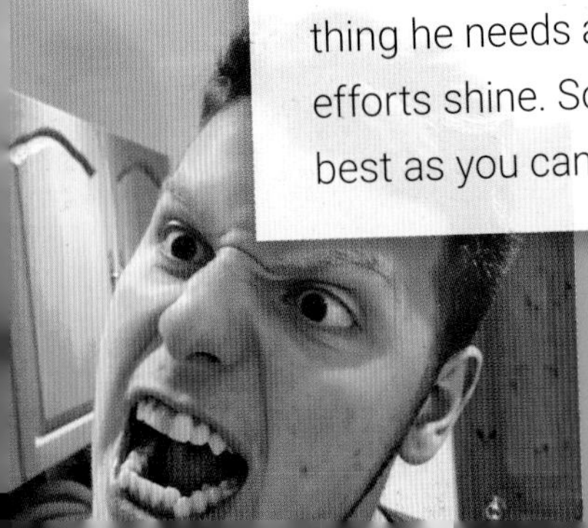

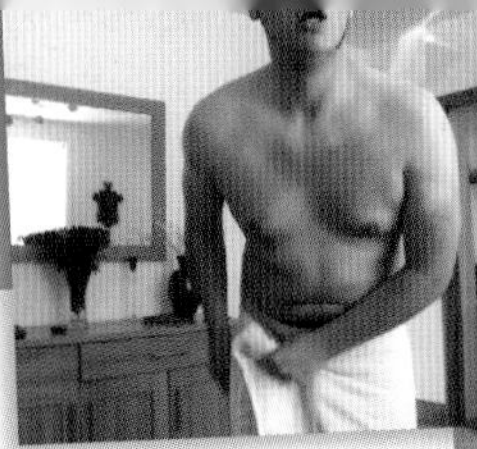

3 PLAN YOUR PRANK

Like a bank heist, if you want things to go well then you need to know exactly what's going to happen and when. So spend time making sure that everything is in place. It might take a while, but the more care and attention you put into the planning stage, the better the outcome.

4 MANAGE THE MOMENT

If I make a mess of sticking an ice-cream cone to Elliot's head, I can hardly ask him to hang around until I get it right. We're talking about a one-off opportunity, so don't mess it up. But even when everything goes right, my work isn't over. I need to calm him down so he doesn't stomp off or break something.

5 PREPARE FOR PAYBACK

If you're going to set up a prank, better brace yourself to be pranked in return. It's only human nature to want to get your own back. All you can do is hope that they make the same effort as you. Weirdly, Elliot's never even tried to get me back. I can only think this is down to his early upbringing in the forest. The day that squirrels and badgers develop a sense of humour, we should all be very afraid.

Chapter 2

HOLIDAY FROM HELL

0

I won't lie. It took me ages to accept that Elliot had become a full-time family member. Even during my time at primary school, I imagined that one day I'd wake up to find my so-called brother gone.

Sadly, that never happened – but I grabbed every chance that came to me. I only have to look back through my first attempt at writing a diary to see how much time I spent trying to lose him from my life.

Dear Direy,

Last night I had a dream that we left Bridgend. Let me tell you, it was TERRIFYING! Ive' lived in this town all my life. I'ts boring here, and I have to make my own fun, but its' safe. Sometimes I wonder if Elliot came from beyond Bridgend. Iv'e heard that many monsters and freaks roam the wilderness over the hills and beyond the motorway.

Let's face it, even though h'es been shaved from nose to tail (Im sure he"s hiding one) Elliot looks like a freak. You only have to look at his black snaggle tooth to know that he ate far too much dirt crawling around in the woods.

I enjoy winding up my so-called brother. It brings out his inner idiot. Take this morning. We went to the supermarket with Mum. She woul'dnt let me leave Elliot outside on a leash with the dogs. He even got to ride in MY seat in the trolley... but not for long.

'Go fetch us some sweeties,' I whispered to him when Mum was at the fish counter. 'Bring me two lemon gobstoppers and you can have some Love Hearts.'

Now, Elliot is mad for Love Hearts. He thinks the massages like 'yo'ur cute' and 'be my baby' are just for him, because lets face it he is STARVED OF AFFECTION!!! As soon as I mention the Love Hearts, he gets out of the trolley and hurries off.

'Wher'es Elliot?' Mum asks when she turns around with her prawns.

'Dunno,' I say with a shrug. 'He just ran away. Not like me. Im' a good boy, are'nt I?'

Mum pats me on the head and looks around.

'Maybe he went this way,' I say when I spot him coming back, and hurry her in the opposite direction. 'Quick, Mum!'

My fun only lasted for five minutes. Eventually, Mum started shouting his name, which was even more embbbarrasing than when she got up and danced at Nana and Grandad's five hundredth wedding anniversary. Suddenly everyone in the supermarket was looking for Elliot. When the security guard found him near the out of date ready meals, Mum burst into tears. I could'nt blame her. I felt just as sad and disappointed. As we headed home, I decided that if I was going to do my family a favour and lose my brother for good then I would have to be sneekierer.

Dear Dairy,

Its' been a long time since I wrote anything. This is because 'Ive been busy! Let me tell you, losing a brother is not easy! Either mum freaks out and everyone starts looking for him, or he finds his own way back to our front door. Honestly, its' like he was'nt raised by wolves but by homing pigeons! I was starting to think I would never get rid of him. Then we went on a family holiday, and I felt sure there would be no way back for him.

'Boys,' said Dad one summers day. 'Pack your bags. Wer'e going on a magical journey.'

'Where!' I asked, while Elliot just stood there and blinked.

'Somewhere exotic' he said, as Mum clapped her hands in delight. 'I know how to treat my family!'

For once, I was so excited I forgot to even try leaving Elliot home alone. With the car packed, Dad drove through the high street and joined a road that led OUT OF TOWN.

'Where are we going?' I asked nervously. Even Elliot looked worried, because we had never been this far from our house. 'What are those green things?'

'Fields,' said Mum, who looked a bit alarmed herself.

A moment later, we passed an army of great beasts. Elliot squeaked in fear.

'Theyr'e cows,' said Dad casually. 'I've read about them in books.'

We drove for many days. OK, maybe half an hour at the most, but it felt like ages. Elliot and I just stared in wonder at the changing scenery, and when we saw the sea our mouths fell open.

'Stop, Dad!' I cried. 'It's the end of the world!'

'Don't be stupid, boyo,' he said as we pulled into the caravan park. 'Its' Porthcawl.'

'Porth-CAWL!' I said, as if repeating what h'ed just said might make it seem more real. 'Did you hear that, Elliot? This is like a dream come true!'

Elliot heard me loud and clear, but just then he was so excited I thought he might wee in his pants. A girl from my class once boasted that she had come here on holiday, but nobody believed her. As Bridgend boys, we were being spoiled here. We were also gonna enjoy it. Not only that, from the moment I learned that the caravan would only fit us all in if Elliot and I slept top to

toe, I swore that my chance to lose my brother for good had finally arrived. Porthcawl was a paradise alright, but I would make it hell for him.

I did'nt rush into anything. For the first few days. Covered in suntan lotion in case the rain clouds ever broke up, we splashed about in the shallows and mucked about with a bucket and spade. We also had some fun burying each other up to our necks in the sand. I thought my turn to bury Elliot would be EXTRA fun. He even helped me dig the hole for him really close to the shoreline. Unfortunately, the tide just went out and took my dream of being an only child with it. My disappointment did'nt last long, however...

'Boys,' said Dad, half way through our once-in-a-lifetime holiday, before showing us a rubber ring he'd just bought. 'Here's a present for being so good.'

Elliot was really excited, but not as much as me. Even before w'ed plonked it in the water, I knew just how that week would end.

'Its' your turn,' I said to Elliot, on the last day after lunch. We were standing in the sea, holding onto the ring as it bobbed about. For once, the sun had cracked through the clouds. It made the drizzle feel quite warm.

'You might even get a tan if you close your eyes and go to sleep.' Elliot gave me a look. I grasped the rope tied to the ring, and promised him that I would hold onto the loop at the end. 'You'lll be fine,' I said. 'What could possibly go wrong?'

Now, we all know Elliot is "special". Not only does he barely speak, he can doze off in seconds. It's like his brain ca'nt handle too much without needing a nap. So, he climbed into the ring, settled back and closed his eyes. A minute later, the bobbing waves had rocked him to sleep. I knew this for sure because he soon started to snore. Honestly, it was loud enough to scare away the seagulls.

It means not even the birds were watching when I pulled out the air plug from the ~~aft port aft~~ back of the ring, pointed him towards the horizon and give it a gentle shove.

'Bon voyage, Elliot!' I said, as the escaping air pushed the ring out to sea. 'Have a nice life as an island castaway. Hope you like the taste of coconuts!'

And with that, I waded back to the shore and waited a while before telling Mum and Dad that my brother had escaped again. He had tried so many times, I just hoped they realised that if he was lost at sea then it would be for the best.

ELLIOT'S DIARY

I'm learning loads about life away from my woodland home. I've got used to sleeping in a bed instead of a burrow. I also prefer chips instead of chestnuts, and a hot bath is a lot better than jumping in an icy stream. Back then, in fact, it was so cold I just didn't bother washing at all.

But the most important thing I've learned is this: if Ben is being really nice to me then I should be nervous. Like, scared-for-my-life nervous. Unfortunately, on holiday, I learned this lesson the hard way.

'Ben?' I muttered when a trickle of cold sea water crossed my ankles and my belly. 'Stop splashing me!'

I was dozing in the rubber ring at the time. My brother had promised to keep me safely in the shallows as I slept, and I'm not the sort to say no to a nap.

'Cut it out!' I cried out again, when the trickle of water became more like a stream. 'I mean it!'

Normally, I would expect to hear him laughing. Mum says it's infectious, but I certainly don't want to catch it. It makes him sound like a girl with her skirt caught in a door. Anyway, this time I couldn't hear him at all. In fact, I couldn't hear anything but the sound of slapping waves and a bubbling noise from the rear of the ring like a never-ending fart.

'Ben?' I opened my eyes, sat up and looked around. That's when I realised I couldn't see land. At the same time, what was left of the deflated float folded in half and sandwiched me. 'Ben, what have you done? BEN!'

This was serious! Miles from shore, clinging to a sinking ring, I bellowed for my brother at the top of my voice. 'HELP ME!' I yelled so many times my voice went hoarse. An oil tanker crossed the horizon, too far away to hear me. After a while, I stopped shouting, and though I was treading water I knew it couldn't last. The wise owls of the woods had taught me how to survive in the wilderness, but this was no forest glade.

Then I thought of the one thing that might just save my life. I hadn't told anyone about this before, because it's a bit embarrassing, but when you grow up in the wild you learn how to talk to the animals. I know, right? If Ben ever realised, he'd be testing me out on dogs and stuff. I can't talk to pets, though. That would be bonkers. But when we watch wildlife shows on the telly I swear I understand what's going on.

Obviously there were no wolves or badgers here, 20 miles off the coast of Wales. I just had to believe that other creatures might hear me. It was a long shot, but I had nothing to lose. Calming myself so that I could do this right, I took a breath ... and began to click and whistle.

By now, the ring had lost all air. I was seriously beginning to sink. As the water rose around my neck, I carried on clacking and trilling, and then gasped when something rose up from the deep and nudged me gently on the arse.

The dolphin, when it surfaced out of the water, looked at me out of one eye. I carried on whistling madly, and it seemed to understand. With a high-pitched squeal, the dolphin circled me to snag the rope loop around its fin. Then it began to pull.

I held on tight as the creature picked up speed, and cried

out when the beach came into view. Finally, I caught sight of my parents on the shoreline. Ben was standing beside them. He'd covered his face with his hands, but I could hear him sobbing his heart out.

'Elliot!' cried Dad, as the dolphin launched me into the breakers before swimming away. 'You're alive!'

As I crawled out of the surf, scowling hard at Ben, he uncovered his face and set his puffy red eyes on me. Straight away, his tears turned to cries of joy and then annoying hoots of laugher.

'Sorry, Bro, but you look like you're wearing a wig made from seaweed!' he cried. 'Hahaha!'

'Ben!' I yelled with murder in mind, and clambered to my feet.

THE COLOUR PURPLE*

Fancy having a go at Elliot yourself? Let's say he got stuffed in a wheelie bin full of paint and pushed down a hill . . .

* AND BROWN, GREEN, BLACK, BLUE, YELLOW . . .

ELLIOT SPITBALL TARGETS

Cut him out, string him up and let the gob fly!

ELLIOT SPITBALL TARGETS

Cut him out, string him up and let the gob fly!

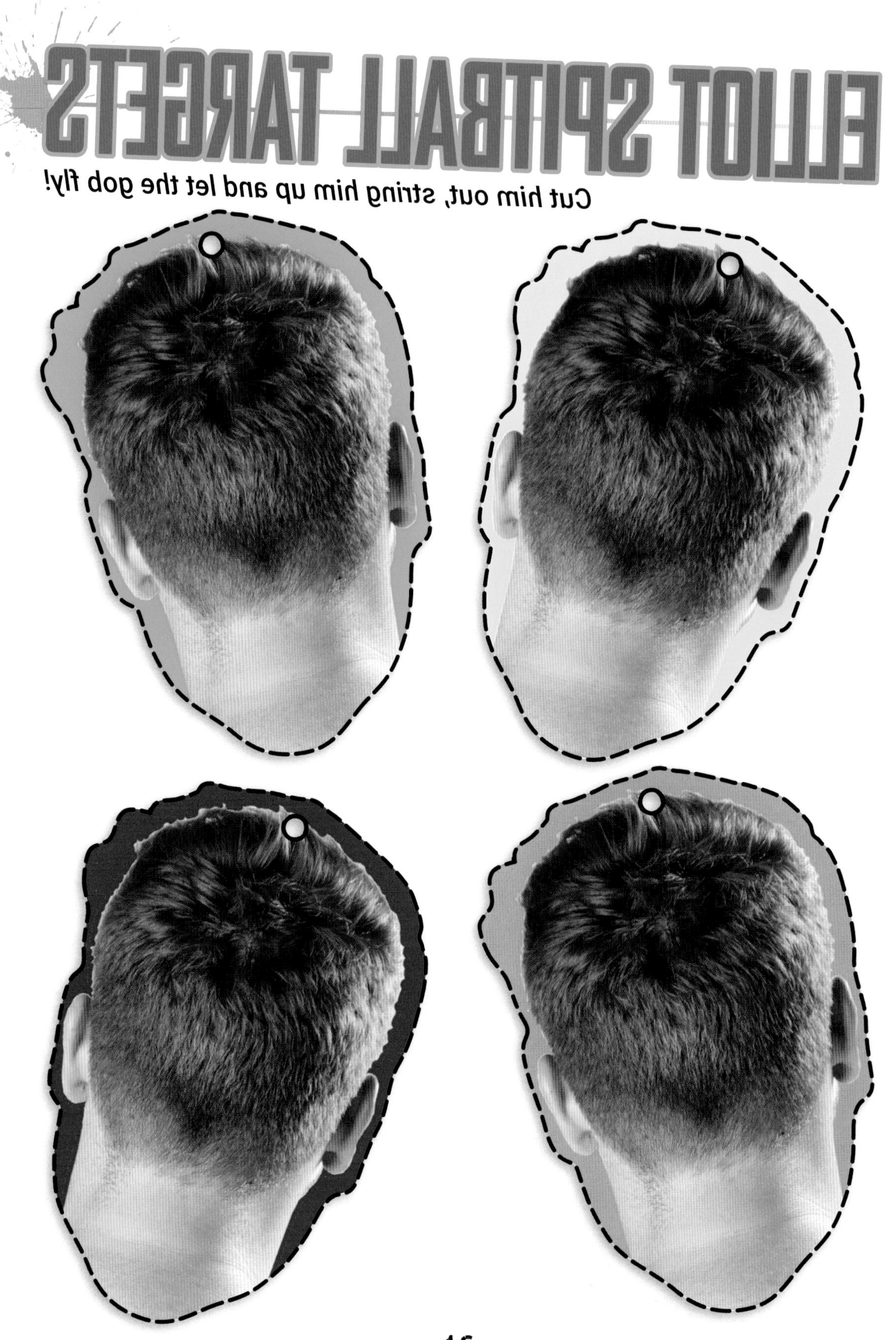

...dot-to-dot

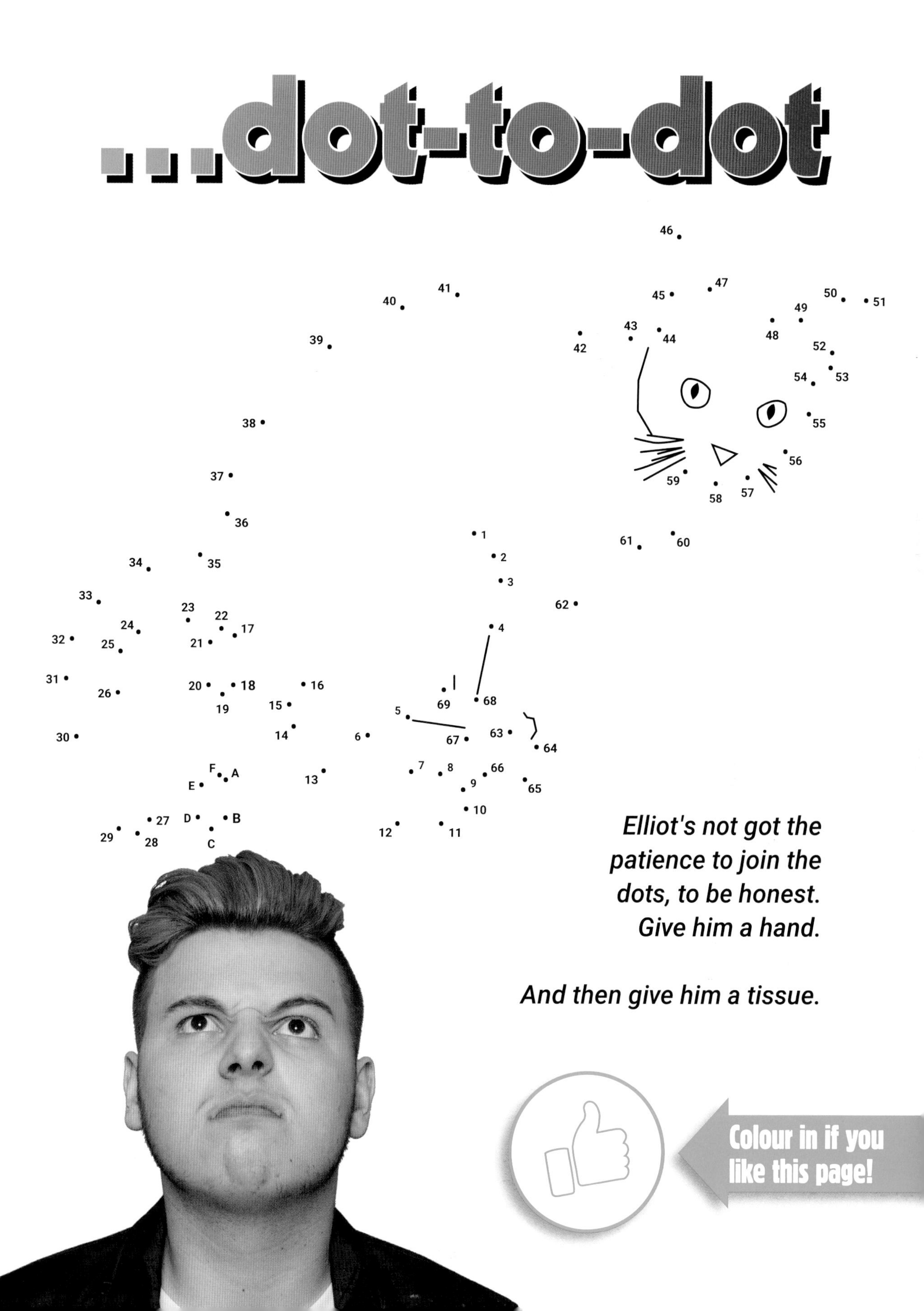

Elliot's not got the patience to join the dots, to be honest. Give him a hand.

And then give him a tissue.

Ben Phillips
@BenPhillipsUK

Q. Ben where do you get all your pranks from?!... A. Up here with my crazy hair 😆

12 Aug 2015

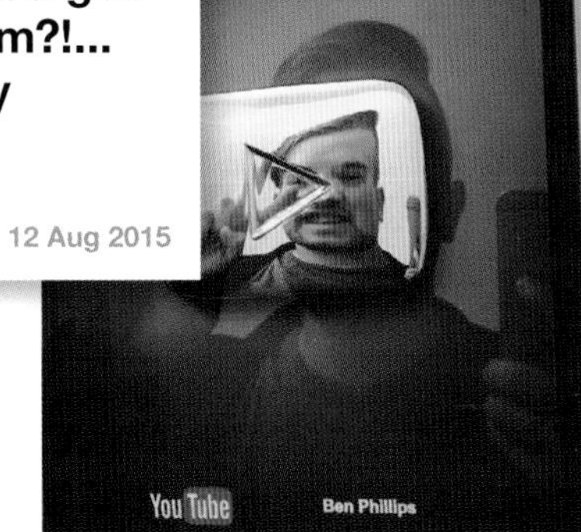

Ben Phillips
@BenPhillipsUK

"I talk a lot of sh*t when I'm drinking, baby"

28 Jul 2016

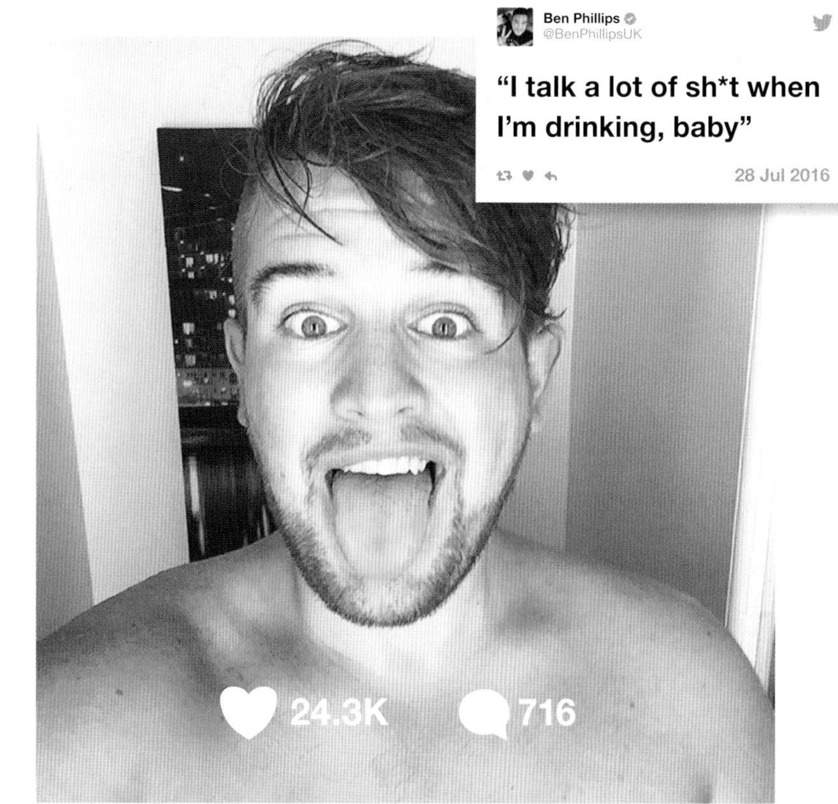

PRANK KIT ESSENTIALS

Snoozers are losers in this game, so if you want to prank with perfection then be ready to act in a heartbeat. Here's what you need to get the job done!

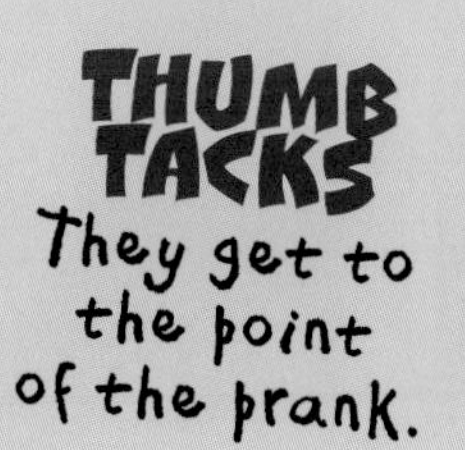

THUMB TACKS

They get to the point of the prank.

PERMANENT PENS

Because those faces don't draw on themselves.

SPECIAL GLUE

(See page 64) – if you need to stick your target to any surface in the known world, this stuff won't let you (or them!) down.

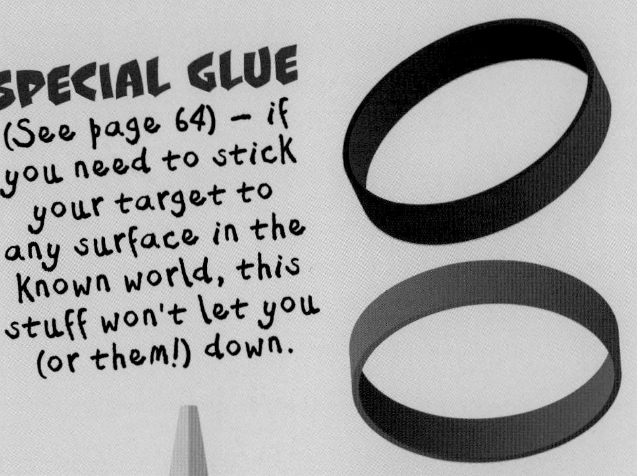

CLINGFILM

From toilet bowls to open doors, this is how you seal the deal.

ITCHING POWDER

Never go anywhere without the old Sprinkle 'n' Scratch.

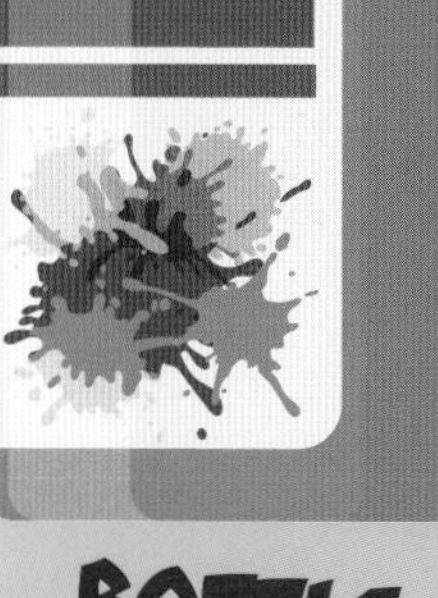

BOTTLE OF RUNNY BUM

A little brown liquid in the right place is always funny, right?

ELASTIC BANDS

Your basic pinger stingers.

Chapter 3

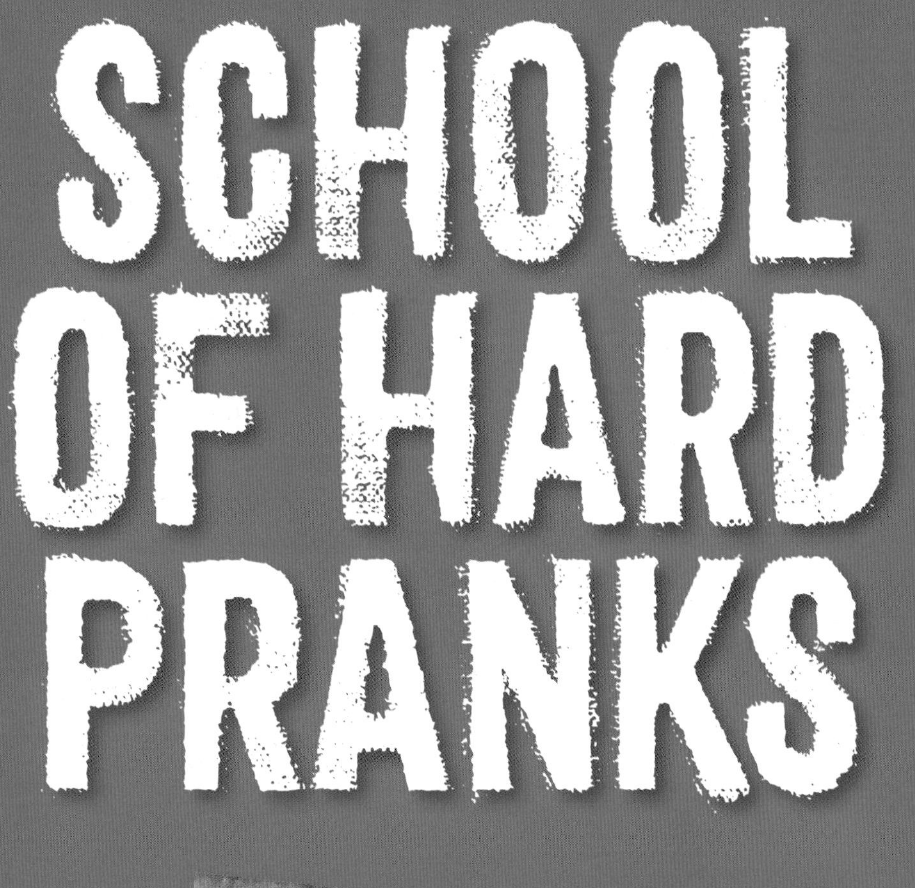

SCHOOL OF HARD PRANKS

0

I learned one very important lesson on that holiday. As Elliot chased me back to the caravan park and forced me to lock myself in the Portaloo, it was clear that my brother could get mad. And when he got mad he got stupid. What's more, because he never once saw the funny side, it drove me on to bigger pranks just to see how far I could go.

From smearing chocolate sauce on Elliot's pyjama bottoms while he slept to replacing the toothpaste in his tube with mustard, I soon showed him home wasn't safe.

Aw, look at that smile...

At the same time, I dreaded the day when Elliot would join me at big school. None of my friends even knew I had a little brother because – let's face it – why would I want to make a big deal of the fact that my parents had been dumb enough to adopt some backwoods urchin who couldn't take a joke and only spoke in gruff sentences like he was desperate to just bark and howl at the moon? So as he got ready for his first day of term, with brand new shoes and school bag, I swore to myself that Elliot would make himself known to everyone without any introduction from me …

So innocent! Didn't expect a thing!

MY DIARY

PRINTED ON MY NEW COMPUTER!

This morning, Mum slipped a good luck note into Elliot's bag. I felt like I needed to add to the nice surprise. Rather than write a card, which was a soppy way of wishing him well, I decided to squeeze the contents of a wet dog-food pouch into the side pocket. It was a nice touch, I thought, zipping up the bag before handing it to my bro.

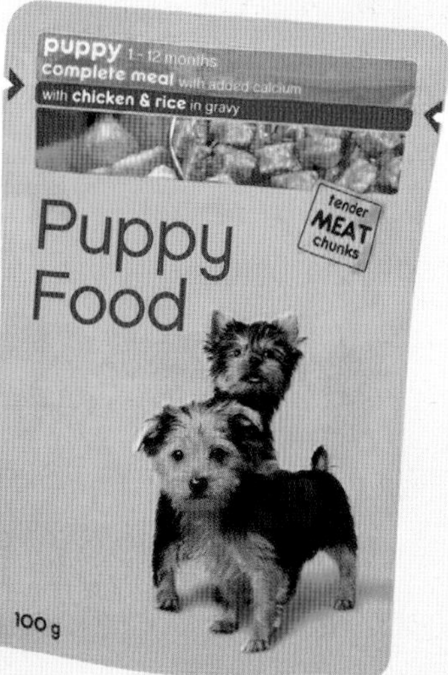

'Have a great time,' I said to him when we arrived at the school gates. 'What a lovely hot day to start the new term!'

Elliot didn't reply. He was too busy wondering why a golden retriever had followed him all the way from the bus stop.

'Shoo,' he said, and flapped his hand at the pooch, but the dog just trailed after him. I watched him go, and sniggered as a collie in a nearby garden wagged its tail excitedly and jumped the fence to be closer to my bro.

I spotted Elliot again just before assembly. It wasn't hard. Before the teachers trooped onto the stage, everyone had turned around in their chairs to see what all the fuss was about.

'Who's that new kid?' asked a girl from my year. 'Isn't it against school rules to bring a pack of dogs with you?'

Poor Elliot! While all the other Year Six newbies were hoping to keep their heads down, make some friends and settle in, my brother stood out for all the wrong reasons. By now, half a dozen hounds had slipped into school to follow him around. Out in the corridor, the headteacher wagged his finger at Elliot, who stared at the floor looking very sorry for himself.

'He can't help it,' I told the girl beside me. 'He's a wild boy, you see?'

It took several members of staff to finally shoo the dogs from the grounds, but Elliot's misery didn't end there. When the school bell rang at the end of the day, they were waiting for him at the gates. The pack had also grown in size. What's more, none of them had bus passes.

'They're not coming on board,' the driver told my brother, who stood on the pavement with his four-legged friends. 'This is a bus, not a boarding kennel.'

I was sitting in the middle row. I would've sat at the back but the bigger boys had got on first. It meant that Elliot could easily hear me laughing along with everyone else.

'See ya, bro!' I shouted through the window as we pulled away, and waved at him from my seat. 'Walkies!'

I hopped off three stops early to buy a lolly from the shop. The bigger boys from the back also bundled off behind me, and then pushed in through the door to get to the freezer cabinet first. There wasn't much left by the time those greedy vultures had finished. So I paid for the Mini Milk and headed out to eat it on a bench in the park. It was so hot that afternoon it felt like the whole of Bridgend would melt away. Some might say that would be a result, but not me. I loved my hometown. Everything I needed was right here. I closed my eyes, lifted my face towards the sun and started to hum a happy tune. I dunno. It was probably a Disney song or something. I just didn't realise anyone was listening until a shadow fell over me.

'Oi! Dopey! Wake up and give us your money!'

I snapped open my eyes to see the bigger boys from the bus. They were standing right in front of me. All of a sudden, I felt a bit stupid sitting there with a red face from the sun and a Mini Milk melting in my hands.

'I . . . I . . . don't have any.'

'You got change from the shop just now,' said the ringleader. 'Hand it over.'

I tell you, this was daylight robbery! I only had about 20p in my pocket, but I decided it was a price worth paying.

'All right,' I said with a sigh, and stood up so I could dig out the change.

The bigger boys watched me closely, then spun around together when a familiar voice piped up behind them . . .

ELLIOT'S DIARY

'Everything all right, Ben? What's going on?'

I was crossing the park when I spotted him. It wasn't a shortcut or anything. I was just hoping that all the dogs trailing after me might spot a Frisbee or a stick and bugger off. Instead, they just followed like I was some kind of pack leader. It was so weird.

'Elliot!' he cried, and for once looked really pleased to see me. It was then I noticed one of the bigger boys surrounding my brother had his hand outstretched like he was expecting some money or something.

'What's with the pooches?' the boy growled.

I'd tried to count how many had joined me, but they kept moving around.

'I just like dogs,' I said, because I didn't want to admit that for some reason I'd attracted them like flies. 'A lot, all right?'

'Elliot,' said Ben, like I ought to keep my mouth shut.

The big boy smirked at me. I didn't like him. I shifted my bag from one shoulder to the other. The dogs followed it with their eyes like they were watching a tennis match. Meanwhile, Ben was looking restless.

'Are you coming?' I asked him.

He stepped forward with a sigh of relief, but the smirking boy stopped him with one hand on his chest.

'Not so fast,' he muttered. 'It's gonna cost if you want to leave.'

I looked at Ben. We both knew that if Nana found out she'd hunt the boys down, and their families, and waste them one by one. While Ben looked like he was scared, I realised one of us would have to speak up. So I cleared my throat to tell the bullies that they really shouldn't mess with us unless they wanted a crazy old lady on their case, only for a long, menacing growl to fill the air.

'Whoa!'

As the boy stepped back, a couple of the dogs moved forwards with their heads low and hackles raised. OK, so one of them looked like Toto from The Wizard of Oz, and a sausage dog was making the most noise, but it was still kind of scary. Not for me, but for this gang of bullies who'd been hoping to take Ben's money. For a moment the boys stood their ground, but all of a sudden I felt like we were in charge here.

Titus, Terror of Bridgend →

'All right, boys. It's time we left.' The ringleader narrowed his eyes and glared at me. 'Just keep your dogs under control!'

'They're not mine,' I protested, but the gang was too busy hurrying away to hear.

I watched the ringleader shoot a glance over his shoulder at us. Ben waved him off with his Mini Milk stick. Then he turned to me. I was just standing there in my brand new uniform, several sizes too big so I'd grow into it. He smiled at me like I'd just saved his skin

'Elliot,' he said, as the dogs went back to drooling at my school bag. 'Let's go home.'

'But I can't bring all these mutts with me,' I complained. 'Mum will go nuts.'

'I can help you there,' he said as we walked down to the path by the river. 'Just empty the pocket of your bag and give it a rinse.'

I looked at the murky water and crinkled my nose. 'It looks a bit dirty, Ben.'

'Yeah, but at least your bag won't smell of dog food any more.'

'Dog food?' I stopped in my tracks, which made Ben chuckle as he walked on 'What have you done?'

'Nothing,' he said, and turned to face me with a shrug. 'Much.'

With my brow furrowed, I swung my bag from my shoulder and unzipped the pocket. I rummaged around inside for a second. Then I screwed up my face. When I pulled out my hand, I found it caked in stinking, heat-stewed chunks of meat in gravy.

'Urrrggghhh!' I cried, like my fingers had just come off.

'Listen, it would've been wrong not to prank you,' Ben tried to explain without grinning too hard. 'Sorry, Bro!'

I shook my hand in disgust, only for gravy to spatter all over my new school trousers. With my brother doubled up with laughter, the dogs went crazy.

'Ben!' I yelled as the pack began to lick my legs in a frenzy. 'BEN!'

Sweet (chilli sauce) dreams, bro...

MAKE YOUR OWN SPECIAL GLUE

I cooked up this recipe by accident one day. It contains some crazy ingredients, but that's what comes from experimentation. The result is a substance with incredible sticking properties, but it only seems to work on a certain type of person – someone raised by woodland creatures, who sleeps as much as he eats, and has a habit of turning psycho when pranked. Like Elliot, it's not big or clever, but I could use it to glue him to the belly of a 747 and be sure he wouldn't come unstuck. Ladies and gentlemen, I give you . . . ***special glue!***

Ingredients

- 5 spoons of flour
- 2 spoons of pig dribble (see Farmer Chris for availability)
- 1 spoon of sand
- A squirt of shaving cream – and another for good measure
- A pinch of chilli powder (dunno if this helps, but wrong not to try)
- 3 drops of spittle from a furious bro (quite hard to capture, but essential as a binding ingredient)
- A handful of crushed nettles
- A dab of hair gel (Elliot always has plenty)
- The essence of sadness and despair (aka a drop of Elliot's aftershave)

Method

Mix the ingredients in a bowl until the stirring spoon stands up in the mixture. Remove and set aside in a darkened room. After three hours, the mixture will have turned to transparent goo. Transfer into a jar and seal with a lid. Your special glue will last for approximately three years, which is the time it takes for your target to grow completely sick of your pranks.

Sticky situations →

Photo-Me

SCHOOL REPORT

Pupil Name: Ben

Teacher's Report: It's always lovely to see Ben's smiling face in the morning! He's such a happy boy, unlike his brother. A bright and creative child, Ben has an active imagination that sometimes requires direction from his teachers to keep him on track. Recently, during a Geography discussion about dormant volcanoes, Ben suggested it might be possible to 'wake one up' by lowering Elliot into the basin on a rope and waiting for him to shout 'Ben!' at the top of his voice.

Ben is a natural at drama direction, though he does have some work ahead in convincing his cast that the forthcoming performance of Hamlet will be enhanced by the installation of a gunk tank over the stage.

Signed: Mr Havali, Head of Year 10

Prospects: A bright and likeable pupil, Ben has the potential to seek a future beyond Bridgend.

(the only way was down, for my bro...)

SCHOOL REPORT

Pupil Name: *Elliot*

Teacher's Report: *Once again, this has been a challenging year for Elliot. On one level he's a pleasant if quiet pupil, but there are times when he can arrive in class in a state of some agitation. He is quick to claim his brother provoked him, but I have never found any reason to believe that Ben is behind it. We are actively working with Elliot to address his anger issues – and tendency to doze off in class – and I hope his work will see an improvement in the coming years.*

To offer a glimmer of hope, Elliot has shown great passion for one subject in particular: food tech. It's not rocket science, but everyone has to start somewhere. Good luck with that, Elliot!

Signed: *Mrs Correll, Head of Year 8*

Prospects: *I have no doubt that Bridgend will suit Elliot for the rest of his burger-flipping life.*

GRILLING GRANDAD

I asked Bridgend's favourite OAP what's really going on inside that rusty old brain, and he didn't tell me to bugger off home. Well, not straight away.

Describe me in five words . . .

Charitable, kind, honest, trustworthy and . . . no, wait, I'm thinking about another grandson. No offence, Ben, but I wouldn't trust you with a pin in a balloon factory.

Describe Elliot in five words . . .

Dickhead. Pervert. Weirdo. Go away.

How did you meet Nana?

In the late 1950's, when times were tough, I was looking for a getaway driver for a job on the local post office . . . is this between us? It isn't? Oh. Well, let's just say that Nana impressed me with her driving skills.

What do you love most about her?

Nana loves a nap, she does. And when she nods off she goes down deep. Nothing will wake her, you see? Honestly, I could fart in her face and she won't even stir. That's a classic, in fact. You can learn a thing or two from an old-school prankster like me.

Sorry Grandad!

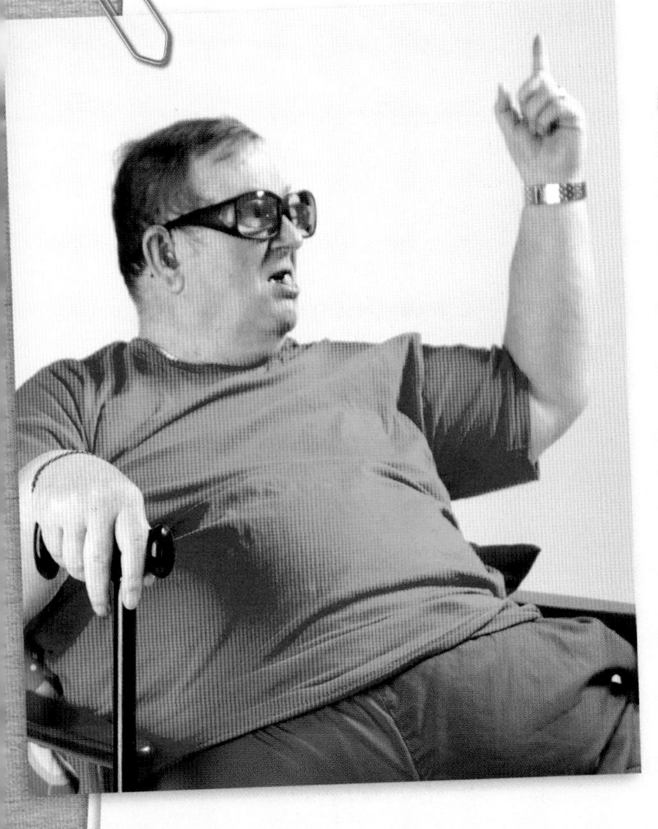

Tell me a secret.

My big glasses with the yellow lenses? They're custom-built so I can see through Nana's dress. Too much information? Well, you asked!

Invent a prank for me to pull on Elliot.

Borrow one of Farmer Chris' horses and next time Elliot takes one of his naps and starts snoring away, dangle a carrot near his mouth and hopefully he'll get his first proper snog.

NO LIMITS WITH NANA

She's old, bold and not afraid to scold (mostly when I've pushed a prank too far). I settled into the comfy seat beside our Nana to find out what's really on her mind.

What's your earliest memory of me?

Well, you weren't very happy when your brother came along. For Elliot's own safety, you came to stay with your grandad and me for a couple of days. You spent most of the time drawing pictures of the new arrival. It seemed to make you really happy, but we didn't keep any because every one just showed your brother shouting at the top of his voice. Now why would he do that, Ben?

What's your earliest memory of Elliot?

A mud-encrusted face with two bright eyes blinking up at me. Nothing much has changed since then, really . . . pervert.

What first attracted you to Grandad?

Well, it wasn't his money. He didn't have two pennies to rub together! But back in the day he was a handsome fellow just like you, Ben.

He even had his own teeth! Snogging him now isn't much fun. If he leans in with his tongue waggling I just pretend to be asleep. Wouldn't you?

Tell me a secret.

Before you were born, I used to work Tuesday nights down at the Cart & Fiddle pub.

Don't you like it, Nana?

No, Ben, not as a barmaid. I was the stripper . . . What? Why have you turned green? If you're about to throw up then get off my carpet!

Invent a prank for me to pull on Elliot.

Tell him I've passed away. Arrange a funeral and watch his face when I rise out of the coffin . . . Too much? Oh, come on, Ben! Call yourself a prankster? Youngsters nowadays are such lightweights!

Chapter 4

WORST BLOG EVER

I'd like to say my first steps online were an overnight success, but that would be a massive lie. The truth, is I fell flat on my face. Still, I tried! It was an exciting time. Instead of tapping out a secret journal that I kept on the family computer, the internet gave me an opportunity to *share* my thoughts and feelings. So before I go any further, I'd like to apologise to you all. I'm sorry, all right? My first efforts were as lame as Elliot's chat-up lines (see page 148), and possibly even more embarrassing.

What went wrong? Well, it's one thing to start a blog, but then you have to fill it. As far as I was concerned, Elliot was the last thing I wanted to include. Why would anyone want to shine the spotlight on their little brother, eh? No, I set my sights on fame and fortune by focusing on a subject close to my heart.

As soon as I set up the page, I ran downstairs to tell the family. Nana and Grandad were visiting at the time, so it took a while to explain everything. Not only were they clueless about what a blog was, I had to spell out how the internet worked. Then I had to lug the computer downstairs because they'd never heard of such a thing. I even had to show that it wasn't coal-fired but used a miracle form of energy called electricity. Then, once I'd brought everyone up to speed, I asked for a special favour.

'Would you follow me, please?'

'Eh?' Grandad looked really confused. 'To where?'

Elliot laughed and tried to cover it up with a cough. Nana asked me to speak up, which is when I gave up trying to persuade the super-oldies. Luckily my parents got the message loud and clear. Well, Mum did at any rate.

'Of course we'll follow you,' she said, and nudged Dad as he snoozed on the sofa beside her. 'Won't we, dear?'

BroBlogSpot, the coolest site in South Wales – Interwebs Searcher

File Edit View Favorites Tools Help

Back Forward Stop Refresh Home Search Favorites History Mail Print

Address http://broblogspot.org/bensdinners/post001/html Go Links

BEN'S DINNERS: A BLOG

Hi food fans. Ever wondered what I have for my tea every night? Well, the wait is over!!! Here you'll find breaking news about each and every meal I eat, plus exclusive pictures of everything from after-school snacks to pudding!!

To celebrate this global debut, we'll begin with a firm favourite in the Ben Phillips household. Yes, it's macaroni cheese!!!!

For the trivia fans, I like three dabs of brown sauce on the side. Yum!

© Ben Phillips. For press enquiries, please contact my mum.

Keywords: Ben Phillips, food blog, exciting, cheese.

Done Internet

Start BroBlogSpot 10:11 PM

And so, with great excitement, I launched my blog to an audience of … three. Not only did Mum sign up and put down Dad's name too, she called Farmer Chris and told him we would be getting our Christmas turkey from the supermarket unless he supported me.

Despite the quiet start, I thought to myself as I clicked refresh a few times to get the traffic counter into double figures, it was only a matter of time before the whole world would be waiting for my next upload. I was made for this moment, I decided, and sat back ready to become an internet sensation.

For a couple of hours that evening, I sat in front of the screen waiting for the comment section to fill. I wondered who would be first. Looking at the time, dawn was breaking in Australia. Surely fans in Sydney and Melbourne would be first in the queue? When that didn't happen, I turned my attention to Japan and then China.

'Hello,' I said at one point, tapping on the computer screen. 'Anyone there?'

As if in response, even though it was due to inactivity, my screensaver kicked in. Sighing to myself, I called downstairs to Mum and Dad.

'Um, my blog is live! Check it out!'

Almost as gorgeous as his sister

I looked at the screen once more, waiting for the moment that my parents spread the news. Then something happened that I didn't expect … My follower count changed, but not in the direction I had hoped.

First it dropped from 3 to 2. Then I lost another one. With both parents giving up on me, I just hoped that Farmer Chris wouldn't lose his sense of humour.

A minute later, my counter returned to zero.

For a moment I stared at my screensaver, feeling all alone, and wondered where I was going wrong. I only lifted my head when Elliot started bellowing my name from his room. I guessed he had discovered the egg I'd hidden under his pillow, but the noise he made about it! He sounded like a stranded cow. Still, it gave me an idea for my next blog entry. Even if it didn't get me an army of followers, the thought made me smile during a dark time in my life.

The next day after school, it came as a surprise to everyone when I offered to make pudding for tea.

'But you never cook,' said Mum.

'There's a first time for everything,' I told her, unpacking the supplies I had picked up on the way home.

Mum watched me warily. I had bought a pack of jelly cubes and a can of cream. To be fair, the recipe I had in mind wasn't rocket science.

'Is this for your blag?' she asked.

'Blog,' I said to correct her, and reminded myself of my new plan to target an audience that didn't rely exclusively on my parents.

We had fish fingers, chips and peas that evening. Elliot stuffed himself as usual, as if raised by pigs, not wolves, while I arranged the food on my plate into a face.

'Are you going to take a picture for your blob?' asked Dad.

Just then, I was so looking forward to pudding that I didn't even bother to correct him. I just ate my food without dropping a single pea, because someone had to teach Elliot a lesson in manners. Afterwards, I even offered to clear the plates from the table myself so we could get on with the main event.

'Well, this a treat,' said Mum, and then gasped when I returned with my creation.

'No,' I said, and set the trifle on the table. '*This* is a treat!'

Technically, it wasn't a trifle. I'd made a jelly in a mould, squirted the entire can of cream over the top and finished it off with a raspberry picked (OK, nicked) from Marge's fruit bush next door.

'Wow!' declared Dad. 'Surely this deserves a picture?'

'It totally does,' I said, and sat down again with my camera in hand.

Elliot sat opposite me with his eyes locked hungrily on the trifle.

'Would you like some?' Mum asked him.

My brother nodded, and then grunted for more after Mum had spooned a little into his bowl. By the time he stopped grunting, Mum had filled his bowl over the brim.

'Maybe you should go first,' I suggested to him, 'seeing as you've taken most of it.'

With his spoon grasped in his fist like a caveman, Elliot caught my gaze and held it for a moment. Even though he looked a bit suspicious all of a sudden, I just beamed back at him and got ready to capture the moment ...

ELLIOT'S DIARY

I don't remember much about life before I joined this family. The wise owls taught me to be smart, sharp and intelligent, but life with Ben has changed all that. I used to spend a lot of time thinking. Now I like watching telly. Can't wait for *Countdown* every afternoon.

Even so, like any wild creature I still have a nose for sniffing out trouble. Usually, it kicks in whenever Ben breaks into that idiot grin, just as he did when I got ready to try his pudding.

'What?' I asked him with my first spoonful parked in front of my mouth. Instead of tasting the trifle, I jabbed it across the table at him. 'Have you put something funny in this, eh? What is it this time, Ben? Laxative? Salt? Hot sauce?'

I had good reason to be nervous. Whenever my brother and I were home alone and he made me a snack, it would burn either my mouth or my bumhole. This time, Ben just looked hurt. Even Mum seemed disappointed with me.

'Elliot, you need to learn to trust your brother,' she told me.

I stared at him across the table. Neither of us blinked. Then our dad sighed so hard that we both turned to face him.

'I'll try it,' he said, and scooped some from my bowl with his spoon.

In silence, Ben and I watched him pop some trifle into his mouth. I waited for a second, expecting steam to shoot from his ears or a stomach gurgle like someone had just pulled the plug from the bath. Instead, our dad smacked his lips. 'Delicious! Elliot, you're missing out!'

RING STINGER

I glanced at Ben one more time, narrowing my eyes. He just sat there with his camera at the ready.

'OK,' I said eventually, holding my spoon like a spear fisherman now. 'I'm going in'

'You certainly are!' laughed Ben

I was all set for it to taste disgusting, but I was wrong. It was pretty good, in fact. But then I was also mistaken in thinking that my brother wanted to picture me sampling his trifle with my tastebuds alone. I just didn't have a chance to react when he reached across the table and pushed my entire face into the plate.

'Sorry, Bro!' he cried, as I looked up at the lens and blinked when the flash went off. 'Now this is what I call the perfect picture for my food blog!' he chuckled, rising from the table.

'Ben?' I called after him, grasping at the nearest napkin. Only we didn't use napkins in our house, so Dad's sleeve had to do. 'BEN!'

That evening, getting ready for bed, I found a red mess all over the front of my pants. It took a moment for me to stop freaking out and realise I hadn't hurt myself. It was the remains of the raspberry Ben had used to top the trifle. I was angry all over again, but didn't even bother hammering on my brother's bedroom door. He'd only laugh and refuse to open up. Instead, I marched downstairs to report what I had found to Mum and Dad.

'This is torture by trifle!' I complained. 'He's getting out of control!'

My parents were sitting together on the sofa. Dad had his laptop from work on his knees. The glow from the screen lit up the grins across their faces.

'What was that, Elliot?' asked Mum without looking up.

'You need to do something about Ben!' I thundered as Dad hit a button on his keyboard.

'I just have,' he said. 'Maybe we were a little hasty in unfollowing your brother's blog.'

'What?'

Grinning now, Dad turned the laptop around. All of a sudden, I found myself face-to-face with my face. Only the one on the screen looked totally surprised and was covered in whipped cream. Once again, I felt as surprised and furious as I had when Ben first shoved my face into that stupid trifle.

'This is gold,' said Mum. 'We've both signed up for more. Even Farmer Chris has emailed to ask what prank Ben might play on you next.'

ARGGHH!!!

DIY time
Super Glue

WANTED
FOR PUBLIC
POOING

He's always going on about how life was better in the old days – before the internet or video games were invented. Sounds like a time that fun forgot, if you ask me, but according to our grandad it was a golden age of pranking. Here's his top four (I was hoping for five, but he fell asleep).

Penny Farthing Fail

Yes, Ben, this is my kind of bike! There's nothing wrong with one massive wheel and a seat on top, with a little wheel at the back to keep things nice and stable. It took a bit of practice, but once you got the hang of it there was nothing that could stop you . . . unless your name was Ellis, our lodger's son.

Now, Ellis was a pale lad who liked to borrow my stuff without asking. He used to take my penny farthing, push it to the top of the hill and wheel down the slope so fast it scattered sheep on the way to market. That's why I decided to teach him a lesson. How? Well, a bike with one big wheel is just asking for a stick to be jammed between the spokes at speed. The next time Ellis took my penny farthing for a spin, I lay in wait at the bottom of the slope and timed it to perfection. The stick went in, causing Ellis to fly from his seat at speed and plunge head-first into the duck pond. He moved out of our house a week later.

Chimney-sweep Stick

OK, Ben, so you think you invented special glue, but you're wrong. Long before your time, I came up with a creation made from coal dust, cow lick and the powdered scabs from Ellis's kneecaps. Ye Olde Pranke Paste, I called it. No idea why we used the

letter 'e' on the end of every word in those days. It was the law, I think. Now, I can't say how it worked, but that stuff could stick!

One time the sweep came round with his boy. Elwood was his name. The kid didn't talk much, and stole a slice of cake from my mam's kitchen table. So I decided to teach him a lesson. While the sweep prepared to send Elwood up the chimney, I carefully brushed the kid's back with Pranke Paste. He nearly caught me out, turning to glare at me, but I just smiled sweetly and wished him luck. Now, my glue took a while to set hard. It meant the boy had plenty of time to crawl up the chimney with his broom before things became a bit 'difficult'.

'Help!' he called out a few minutes later, having managed to pop his head out of the chimney top. 'I'm stuck fast!'

The sweep had never known anything like it. He clambered up onto the roof and tugged at Elwood, but he wouldn't budge. In the end, I saved the day using nothing but some kindling wood and a match. Unlike your new-fangled glues, my Pranke Paste melted at temperature. So the boy escaped with a scorched backside and refused Mam's offer of a slice of cake to make it better. He never swept our chimney again.

Bed Sledge

Long ago, it used to snow at Christmas without fail. We'd wake up to find a white blanket had settled over Bridgend. It wasn't exactly picture-postcard. The place was a pigsty back then as well, but we made the most of what we had. And what better way to enjoy the snow than by going sledging?

You can forget about those plastic efforts that kids use nowadays. Our sledges were built from solid timber, and if you didn't bail out at the bottom of the hill then chances are you'd break some bones. It was hardcore, Ben. We knew how to play. And prank, now I think about it.

Take the boy next door. Elyas was his name. Everyone called him Smelly Arse, of course, but he never complained. He was barely awake enough to notice. Honestly, that boy liked to sleep even more than your brother. He wasn't bothered about the snow, but I didn't want him to miss out. So, together with my old mates, we crept into his room while he was sleeping, wheeled his bed out of the house and pushed it up the hill. At the top, we lined him up for a good run, and then woke him up with a snowball in the face. A second later, poor Elyas is bombing down the slope and yelling for mercy at the top of his voice. Luckily his bed was built from wrought iron. It only buckled a bit on smashing into the wall at the bottom. Elyas was fine, apart from being knocked unconscious, which was quite handy actually. It meant we could wheel him back to his bedroom before he woke up with a headache and the memory of the worst dream ever. Happy days!

Tin Tub Trauma

During the war, when times were tough, Mam took in an evacuee from the city (wherever that was). She was an odd girl with a strange quiff hairdo and a brain that seemed to tick like a clock that needed winding. Ellie Yurt was her name, and though she wasn't quick-witted she had no problem putting me in my place.

'You smell a little . . . rank.'

That was the first thing she said to me. Rude!

'What do you expect?' I asked her, and pointed at the old tin tub out in the yard. 'I'm always last in the family queue for a wash. I get out dirtier than when I climb in!'

Ellie Yurt crinkled her nose.

'That queue just got one person longer,' she sniffed. 'And she's in front of you.'

Now, I wasn't having that. After a week of waiting an extra half an hour for a scrub, shivering in my pants while our guest splashed about in the suds, I decided enough was enough. The next day, I headed for the pond with my fishing rod, and stayed there until I'd caught the biggest pike in the water. It was a beast. With a set of crooked teeth. Really. You wouldn't want to mess. Anyway, I took that mighty pond dweller home in a bucket, and sloshed it into the tub ahead of Ellie's turn. Even then, the water was so dirty you couldn't see a thing. I watched a couple of ripples carve across the surface, smiled to myself and then crouched behind the outside toilet (I know, right? Told you times were tough!).

From my hiding place, I watched Ellie settle in the tub. She rested her head back, closed her eyes against the drizzle and began to hum to herself. It was a happy-sounding ditty, until it ended in a gasp and then a scream.

'M . . . m . . . MONSTER!' she yelled, scrambling from the tub. 'ARGGHHHH!'

Mam came running from the kitchen, but by then poor Ellie was gone. She didn't even stop to say goodbye when I emerged with tears of laughter streaming down my cheeks. Nor did she bother leaving through the back gate. Instead, we were left looking at an Ellie-shaped hole in the fence, while I denied having anything to do with the giant creature of the deep swimming circles in the tub. That pike didn't bother me when it came to taking my bath. We had bonded over a prank pulled to perfection, and by the time I slipped it back into the pond that evening I'd even given it a name.

'Goodbye, Elliot,' I said, watching it drop to the bottom to feed. 'Thanks for the laughs, pal.'

WHY YOU RECORDING?

WHO TOUCHED MY NIPPLES?

LOOK AT ME! IT'S NOT A GOOD LOOK!

YOU'RE ALL JUST FREAKS

Google says I got fleas!

This is your fault. All you do is cause me problems

Stop exposing my body on the internet!

HAPPY BIRTHDAY, YOU OLD BAT

NOW THEY'RE GOING TO CALL ME SLAGGY GILES

THEY DON'T CALL ME BIG **** GILES FOR NOTHING

DON'T YOU LAUGH.

DON'T. YOU. LAUGH.

WHERE'S HIS BLOODY TEETH?

I'M SINGLE AS A PRINGLE, OK?

CLINIC. IT'S A CLINIC JOB.

I'VE NEVER HEARD OF ONE. A SCROTUM? WHAT THE HELL IS THAT?

I DIDN'T PUT DYNAMITE IN THE CAKE!

THERE'S ONE PIECE OF GOOD NEWS, BEN. IT'S COME OFF. THE BAD NEWS? IT'S BURNT MY BLOODY SKIN.

I'M WEIRD AS IT IS

Bon Appe-Prank

Everyone has to eat, but then there's Elliot. He doesn't just love his food. It's pretty much the only reason why he wakes. From an early age, I learned that anyone or anything that gets in the way of my bro and his grub will bring out the very worst in him. So it's only right that I mess with his meal whenever I get the chance. Check out my best three servings …

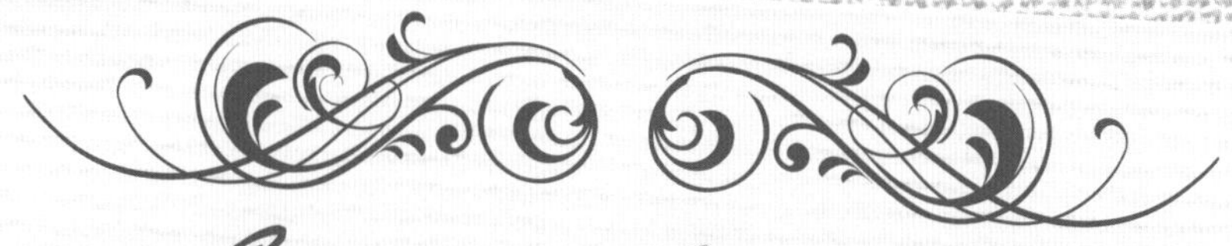

Currant Bunnies

If there's one thing that looks like a raisin, it's rabbit poo. All it takes is a quick sweep of the hutch and you have a whole serving. Next, pick up a currant bun from the baker's and nip out all the nice bits. Then replace with those demon droppings and present on a plate with a nice cuppa. After that, my advice is to step well back and cover your ears before your target explodes.

Smoothie Surprise

Elliot's favourite new toy is his smoothie-maker. You can chuck anything in and the blades blitz it to a pulp. It's perfect for fruit, veg . . . and disguising the most stomach-churning ingredients. Only last week, I fixed a Kiwi Mince Surprise for my bro. He drank the whole thing in one go before I revealed what had given it such 'body'. Then he set the record for projectile vomiting. Result!

Moggy Meatballs

Even if you don't own a pet, if you want to be a master prankster then keep a stock of wet cat-food handy. Why? Because for me there's nothing funnier than watching Elliot tuck into the meal I can make by rolling the contents of one pouch (or two) into meatballs. All I have to do then is drop them into a nice tomato sauce and simmer for 20 minutes. My bro has no idea what he's shovelling into his mouth until the taste kicks in. Then I stand well back as he coughs it back up while trying to yell my name.

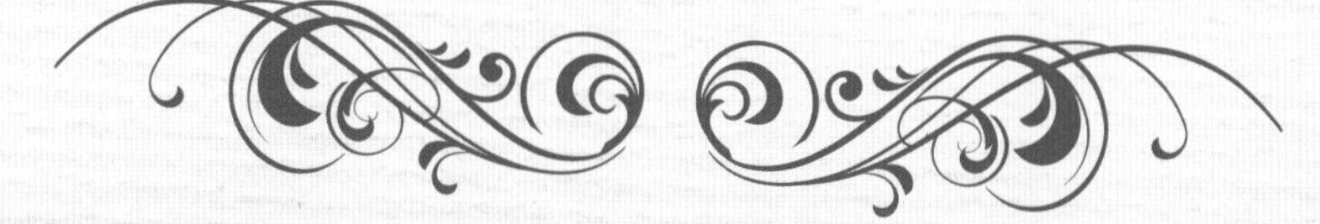

Chapter 5

THE SOCIAL MEDIA BUG BITES

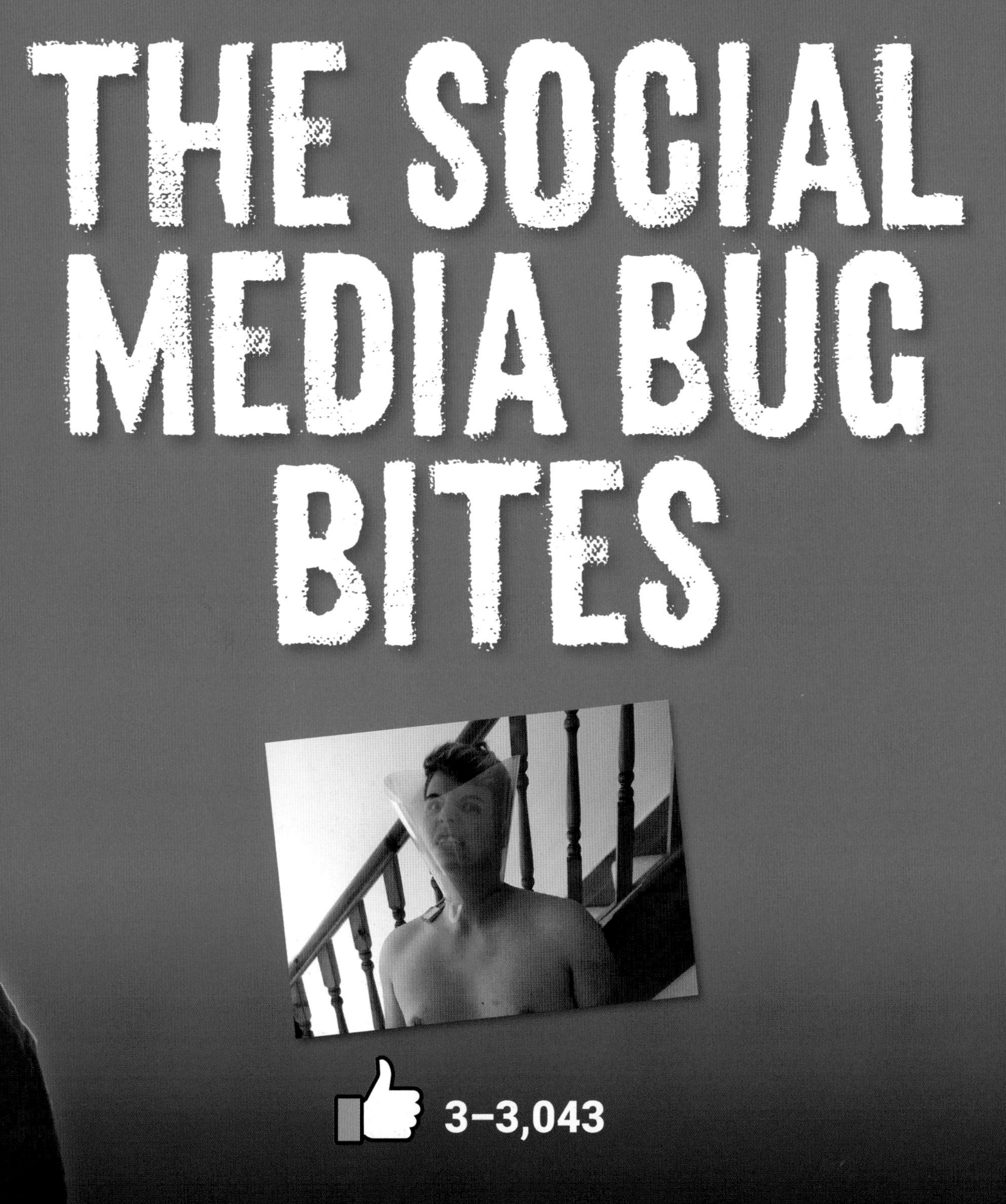

One year after I relaunched my food blog, I hit a milestone. With regular posts of my brother wearing a pudding all over his face, my follower count reached double figures. At last, I had a fandom that ran into the tens of … well, tens.

I could count on a couple of friends from my class to check out my work, along with Farmer Chris and his strange sister, Georgina. I was a little bit alarmed when she started following me. I think it's the beard that I find a bit odd. Then again, she was the first to follow me after a dry patch of several months, so I was happy enough. Unlike Elliot, who was frankly steaming when I told him that slowly but surely my blog was growing in popularity.

'This has gone too far!' he ranted. 'Next you'll be telling me it's up to … *15*!'

It was a figure that might've struck Elliot as his worst nightmare. For me, it was a target I just had to reach. The question was how to get it to such a staggering number. I couldn't just rely on my mates, my mum and dad, a farmer and his freaky sibling to spread the word. I had to go viral on my own.

And so, driven by ambition, I set my sights on transforming my blog into a *monster*. Whatever it took, I was determined to prove to Elliot that I could shoot for the impossible, and increase my follower count by another five.

But before I could even hope to bring anyone else to my blog, I would need bigger and better equipment. At the time, technology was moving on in leaps and bounds. It meant our family computer was starting to creak, and so I hit the online auctions, hoping to bag a cheap laptop. The trouble was, most of my pocket money went on pudding ingredients so I could pie Elliot on camera. As a result, I soon found myself outbid. For several days and nights I saw my hopes rise and fall as bidders with more cash forced me out of each auction. Just as I was beginning to think that I should give up blogging altogether, I came across an offer that seemed too good to be true. Not only did this piece of kit look the business, the auction only had five minutes left and NOBODY HAD PLACED A BID!

How could I resist? I read through the details twice to be sure I hadn't missed anything, and then rushed to put in a bid. This laptop was such a bargain that I still had money left to stay in the game if anyone else came in with a higher price. With my eyes fixed on the screen, and my mouse pointer hovering over the BID AGAIN button, I watched the counter tick towards the moment the auction closed. I felt sure that someone, somewhere, would take one look at a laptop that was practically being given away, but nobody else put in a bid but me.

Holding my breath, I watched the counter tick through the final seconds. Then, with a bleep, the auction status switched to 'FINISHED'. At the same time, I received an email to confirm my item was on the way.

'That was quick,' I said to myself, thrilled that I'd won. 'The seller must be really keen to get rid of it.'

Now, I wasn't stupid. Not like Elliot when he gets mad. I realised that a laptop this cheap wouldn't be perfect. Even the seller had been clear that it contained a bug of some sort, but I was ready. As the days passed by, I went out and bought an anti-virus thing. As soon as my new machine arrived, I would clean it up and get on with chasing my wildest ambition: *15 followers!*

Annoyingly, when that day arrived, it was Elliot who opened the door to the postman and signed for the package.

'What is it?' he asked when I snatched the box from him.

'A surprise,' I said. 'All I can say is that it's going to change my life.'

'What about mine?' he asked as I climbed the stairs to my room.

I turned at the top to face him.

'You're just fine as you are,' I said. 'In fact, it's really important that you never change.'

'What does that mean?'

'Oh, nothing,' I called back as I pushed on for my bedroom. 'Hey, Elliot, I might make a special pudding later. Hope you're hungry!'

'Always,' he said before I closed my door behind me, which was good to hear.

My laptop had come half way round the world, so it was no surprise to find the packaging carefully taped. It took me ages to get the thing open, which made me wonder if the sender was trying to protect the contents or me. When I finally opened the box flaps, I found myself looking down at a load of polystyrene beans with a card on top that read:

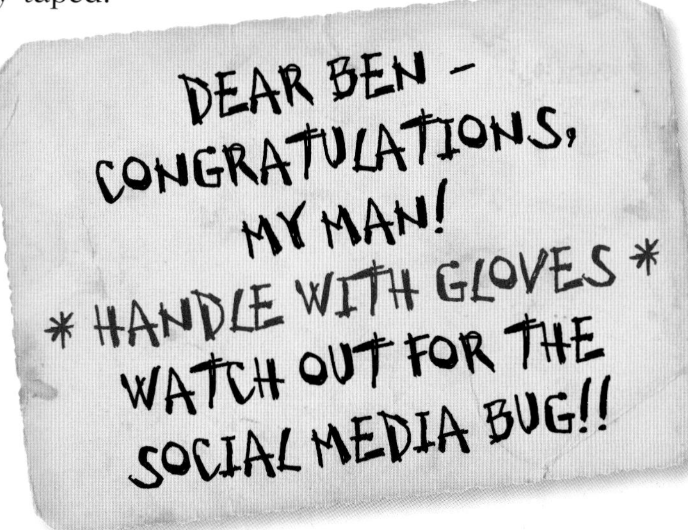

'Gloves?' I said to myself, and dived in anyway. 'A laptop is hardly going to bite!'

I drove my hands deep into the polystyrene filling, and grinned to myself when I found what I had been waiting for all this time. The laptop casing felt cool to touch. I took it gently between my fingertips and began to lift it to the surface. If it contained any kind of software bug, I would zap it with my anti-virus. In my mind, I assumed it was some string of rogue code just waiting to be crushed. What I never imagined was that the bug in question might be REAL – like the kind of thing you hear about people finding in a crate of bananas – with eight scuttling legs and two little fangs set to sink into the back of my hand and make me yelp out loud.

'Ouch!' I cried and yanked my hand from the box. Whatever had just bitten me flew across the room, hit the wall and sparked like a cap gun. I rushed across to see what it was. The thing was no bigger than a fly, all curled up on the carpet with a wisp of smoke rising from its body. Carefully, using a pencil tip and a sheet of paper, I lifted it up for a closer look.

At the same time, Elliot knocked at my door.

'Ben? BEN! Why you shouting?'

'No reason! Now go away,' I called back, but he let himself in anyway.

My brother found me on my knees, peering at the bug on the sheet of paper.

'If that's a nit,' he said, 'it didn't come from me. Mum shampoos my hair with special stuff every Sunday.'

'It's no nit,' I muttered as he dropped down beside me. 'It's ... mechanical.'

Barely had the word left my lips when the doomed bug on the sheet of paper popped apart as if a miniature spring had just broken inside. Elliot and I glanced at one another in shock and fear.

'Looks broken,' he said in a whisper. 'Is it too late to get your money back?'

'I didn't buy the bug,' I told him. Carefully, I pulled the laptop out of the box. It looked as good as new. '*This* is what I wanted!'

Elliot looked confused.

'Since when did bug houses look like laptops?'

Before I could spell it all out to him, Elliot turned his attention to the bite on the back of my hand.

'It's just a scratch,' I said, even though it was beginning to throb.

'Are you feeling OK?' he asked. 'You look like you're going to be sick all over the carpet. As long it's your carpet, then go ahead. Don't you dare make a run for my room and do it there!'

'I'm fine.' I rose to my feet, a little shocked perhaps, and that's when my vision turned blurry ...

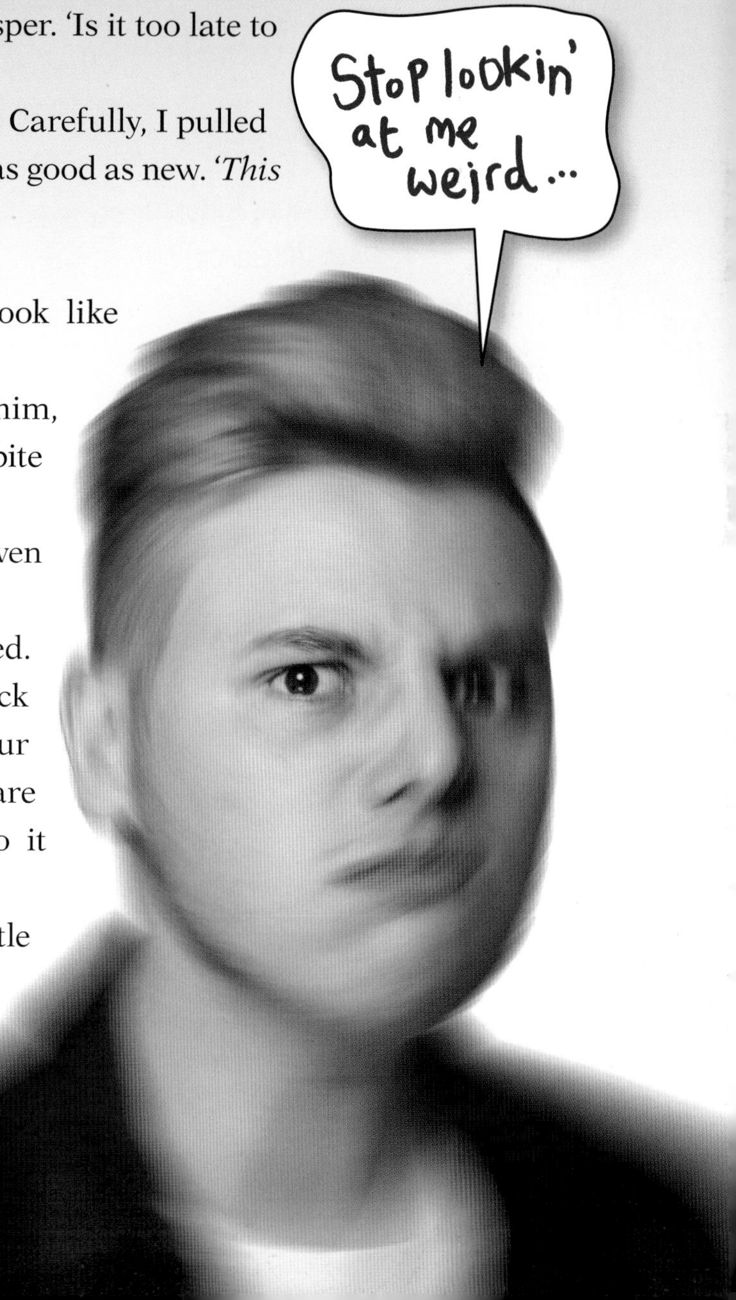

ELLIOT'S DIARY

There was nobody else in the house but us. Mum and Dad had gone to see Farmer Chris to help him shear Georgina's beard. Nana and Grandad only lived round the corner, but just then I didn't think they'd be much help.

'Ben?' I said as he began to shiver and then bleep and trill like an old-school computer modem. 'You're beginning to scare me a bit here.'

With the bite on his hand turning an angry red, my brother's eyes rolled into the back of his head. Then he started babbling. It sounded like code a nerd might program into a computer.

'Stay away from me, Elliot!' he yelled, breaking out of the bonkers code in a voice that definitely didn't belong to him. 'I'm going viral!'

'Is that bad?' I asked.

'It's awesome!' he cried in a demented kind of way. 'Bro, I feel a connection going on like you wouldn't believe!'

Not only was Ben grinning and gurning, his entire body seemed to glitch and pixelate. I even wondered whether he was experiencing some kind of weird software update. Then his eyes dropped back in their sockets. They had gone as wide as webcam lenses, and that's when things turned really crazy.

At this point, as Ben started doing a kind of high-speed dad dance, I decided it might be safest if I stepped outside and jammed the bedroom door shut.

'Whatever transformation is going on, maybe you can let me know when it's complete,' I whispered, and backed out of the room. 'Oh, and don't die or anything.'

Outside, with my back against Ben's bedroom door, I closed my eyes and wished that he would stop ranting, howling and singing. Especially the singing. It was terrible and so out of tune it felt like my ears might start bleeding. Then he started yelling 'LIKE! SHARE! FOLLOW! SUBSCRIBE!' and something about a dream to entertain people around the globe. I thought about calling 999, but at that moment I was so panicked I couldn't remember the number. All I could do was hope he'd stop hurling stuff about in there and make it through in one piece.

Finally, after what felt like hours (but turned out to be a couple of minutes) all the banging and the shouting died down. A moment's silence followed. Then I heard Ben's voice behind the door.

'Elliot? Let me out.'

I was so relieved to hear him sound kind of normal that I didn't think about my own safety. I stopped blocking the door, turned around and opened it up. My brother stood before me. The room behind him was completely trashed. His new laptop was perched on the side of his wardrobe, which lay toppled across the carpet. As for Ben, he looked just the same but completely changed. There was a glint in his eyes like little torchlights were shining through.

'Is it over?' I asked nervously.

Ben invited me to check out his new laptop. I recognised his blog straight away, and sighed at the image of my face just moments after it had been pushed into a soggy-bottomed chocolate cake. I glanced at Ben, because I had seen enough of this image. Proudly, he pointed at the follower count. It took me a second to realise that the blurred bar at the bottom of the screen was, in fact, spinning digits. I watched it whizz straight past the 3,000 mark and then carry on climbing.

'I had hoped we might hit 15 followers,' he said. 'Now I've been bitten by the social media bug and my blog is drawing people like moths to a flame.'

'That's a lot of moths,' I said as the numbers continued to build. Then I looked around to face him. 'Is it going to stop soon?'

'Elliot,' he declared and spread his hands wide. 'Whatever's going on here has only just begun . . . Sorry, Bro!'

'For what?' I asked.

'I dunno,' he shrugged. 'But I have a feeling it isn't going to end well for one of us.'

ANYWHERE

ANYWHERE

ANYWHERE

ON THE FENCE WITH FARMER CHRIS

Some pranks are bigger than others, and that's when I need a spare pair of hands to help me. Farmer Chris isn't afraid of heavy lifting (usually Elliot when he's sleeping), and owns plenty of land so we can leave my bro stranded. Best of all, Chris can call upon all kinds of livestock, from cows to sheep and pigs, and more manure than anyone could ever wish for.

How is life, Farmer Chris?

Mustn't grumble, Ben. Right now, the weather is good, the lambs are skipping in the meadow and Elliot is asleep in the caravan down on the soggy bottom paddock.

What's he doing there?

He rents it from me. £15 a week I charge him. Every now and then, he climbs in there for a kip to get away from you. Sometimes, when he's feeling the need to get in touch with nature, he pitches his tent by the river and kips in there.

Prince, the gay horse →

Should we wake him?

There's no need. I've already filled the caravan with six of my finest sheep. Any moment now, he'll open one eye and start to bellow.

No offence, Farmer Chris, but I call the shots when it comes to the pranks.

Yeah, but he owes me rent, see?

Giddy up, partner!

Fair enough. So, what's your favourite-ever prank?

Pulling Elliot through the manure by a rope tied to a tractor was pretty funny, though he made a mess of my pile. Took me all afternoon to shovel it back into place.

Would you ever employ my brother to work on the farm?

As what? I already own a donkey, Ben. I suppose he might be able to stack hay bales in the summer, but then chances are he'll build a bed and fall asleep for the rest of the day. So, no, to answer your question. Elliot's good for nothing but pranking. Now, get off my land!

Which one to put on Elliot's dating profile next? Hmm...

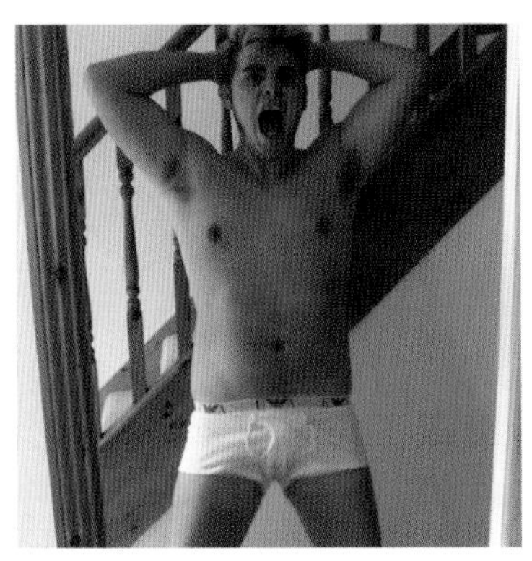

35.1K
302

HYSBYSIAD TÂL

Keep Smiling

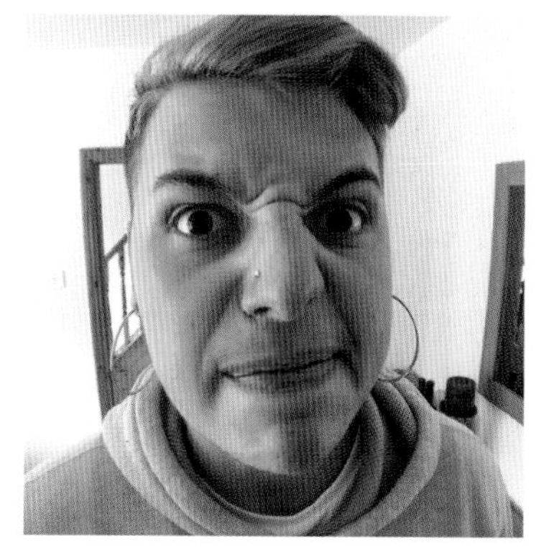

BEN & ELLIOT'S EPIC JOURNEY

Elliot has a dream. One day, he says, he'd like to find out what lies beyond the Severn bridge. This is the link between Wales and the badlands (AKA England). As boys from Bridgend, we've heard all the stories about monsters and stuff that lurk the hills, but it's time we found out for ourselves what lies in wait. Being a kind and loving type of brother, I decided to make his dream come true.

8.00am

Elliot wakes up with his usual morning shower (a bucket of cold water thrown by me). Once he stops ranting, I tell him to pack a bag.

'Bro, it's time for an adventure!'

'Are we going down the sports centre?' he asks.

I tell him to think big, and head downstairs to sort breakfast. We need feeding properly before setting off, which means double eggs for me and porridge for Elliot (with a pinch of chilli pepper, obvs).

9.15am

The time has come. Elliot has packed a suitcase, strapped together with tape, and I'm travelling light with just a rucksack.

'What've you got in there?' I ask as he struggles to squeeze through the front door.

'The usual,' says Elliot. 'Food, pants, Speedos.'

'For swimming?

'No, for pole dancing,' he sneers. 'Of course for swimming! We're going somewhere hot, right?'

I'll admit, I've no idea what lies beyond the bridge, but I tell him it's fine to pack his skimpy swimwear. Mainly because I've cut a hole in them so his 'bro globes' can get some fresh air.

10.22am

The bus is late. We're wet from all the rain, but I'm excited! Elliot just looks as nervous as he did when we visited Porthcawl.

'It'll be just the same,' I tell him. 'But different.'

'Different how?' he asks

'Well, we're going east instead of west.'

Immediately, Elliot worries that we might die from frostbite.

'We'll need warm coats,' he says. 'Or animal skins.'

'Elliot, this isn't *Game of Thrones*. Calm down.'

We travel for the next half an hour in silence. Elliot looks restless. I'm just glad I packed a thermal vest.

10.57am

We arrive at the bridge! This is it. I can't believe it, and Elliot is just as gobsmacked. We jump off the bus and check out the view across the water. Then Elliot turns to me with his eyes as wide as saucers.

'Norman,' he says, like I should've remembered next door's cat. 'I left my bedroom window open. You know what that means?'

'Well, it won't smell of stale farts while you're away.'

'My bed!' he says, sounding really stressed now. 'He'll poo all over it!'

By now, it's clear to me that Elliot is panicking. We're due to jump on the next bus to take us across the bridge, but my bro is pacing up and down like he can't believe what an idiot he's been. I decide to take control of the situation.

'Do you want to see the world or not?' I ask, grasping him by the shoulders. 'Elliot, this is our chance to push ourselves, and see things we've only ever imagined. Starting with . . . England.'

My brother holds my gaze. Then he stares across the water with me. In the distance, I see smoke. It's probably just a factory, but looks a lot like a volcano. I look back at Elliot, who's cacking himself now.

'All these things that we could discover,' he says. 'Can't we just see it on the telly?'

'Do you mean like . . . documentaries?' I ask.

We hear a groan from across the water just then. I'm pretty sure it's a ship's foghorn, but I'll be up-front here: it does sound pretty dragon-like.

'If we turn back now,' he suggests, 'we could spend the day on the sofa. Y'know? Educationing ourselves . . .'

'Bro,' I say, and breathe out long and hard. 'Don't ever suggest we go on a journey again. I knew it was a stupid idea from the start.'

14.09pm

Elliot and I stumble through the front door, tired and hungry. We left as two local boys, and returned as, well, the same really.

'I need a lie-down,' says Elliot with a sigh.

'Good idea,' I say, and head through to the kitchen. 'I'll even fix us something to eat that isn't going to leave you needing to be near a loo for the next 24 hours.'

'Do you mean that?' says Elliot.

'Of course!' I say. 'Diarrhoea is funny, but there's a time and a place for pranks.'

Elliot nods, smiling at me.

'Thanks, bro,' he says, before heading upstairs.

'My pleasure,' I tell him. 'You can count on me.'

I mean it, too. Elliot might be made for winding up, but after an ordeal like the one we've just been through, it's only right that we enjoy a calm, quiet moment as brothers. Unfortunately, that peace and quiet comes to an end within seconds.

'I'm going to kill him!' I hear Elliot cry.

I turn around in time to see my brother stomping towards the back door.

'What's the matter?' I ask as he grabs the key to the shed.

'Norman has been here!' he rants at me, and I realise what this means. 'Don't stand in my way, Ben!'

'But what do you want from the shed?' I call after him.

'The hedge trimmer!'

'Why?'

I don't need to detail what Elliot threatens to do to the poor cat next door. As I rush to grab my camera in case he shreds his own trousers in a rage, all I can say is that it's good to be home.

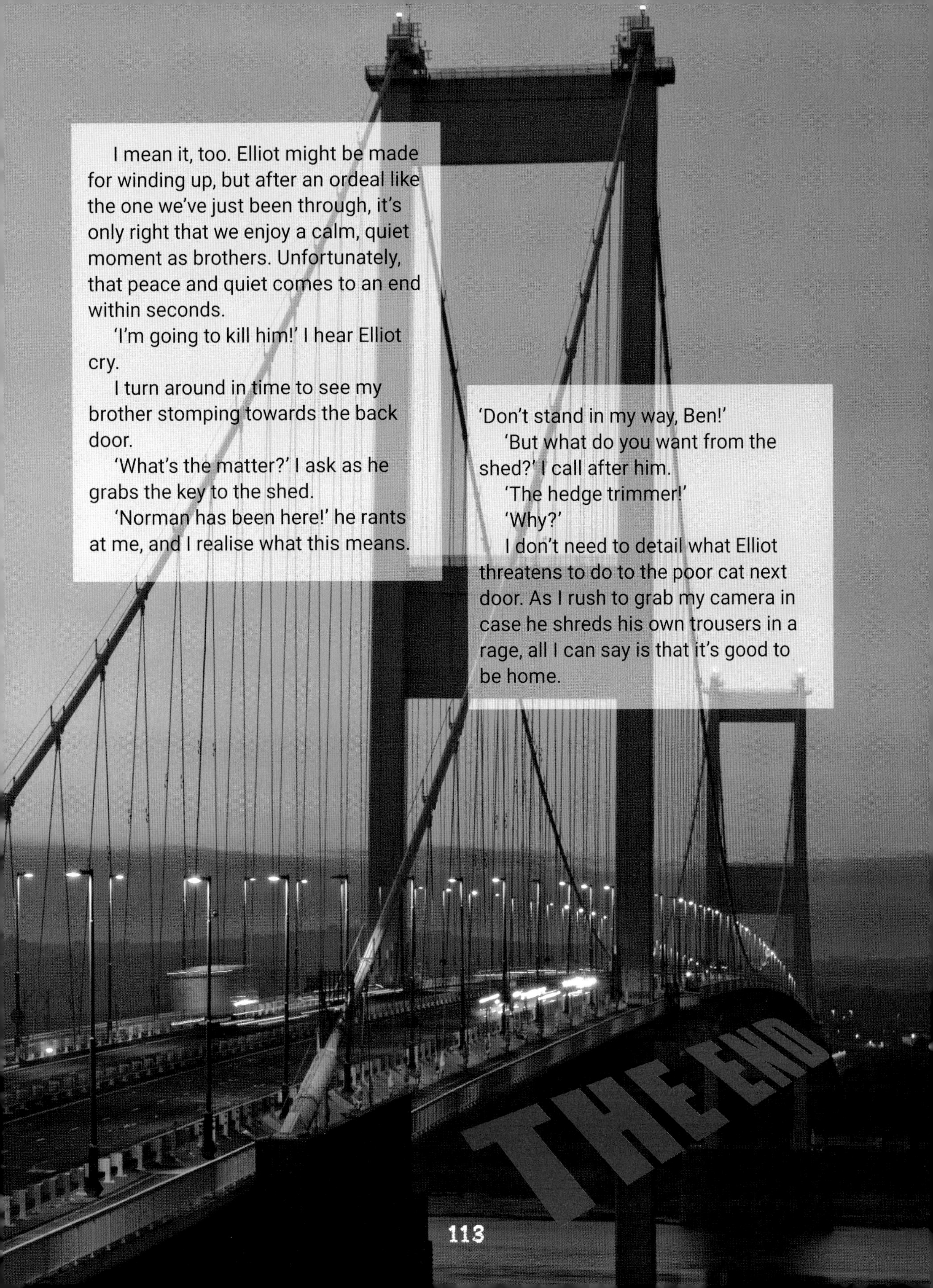

THE END

Chapter 6

VIRAL LIFE

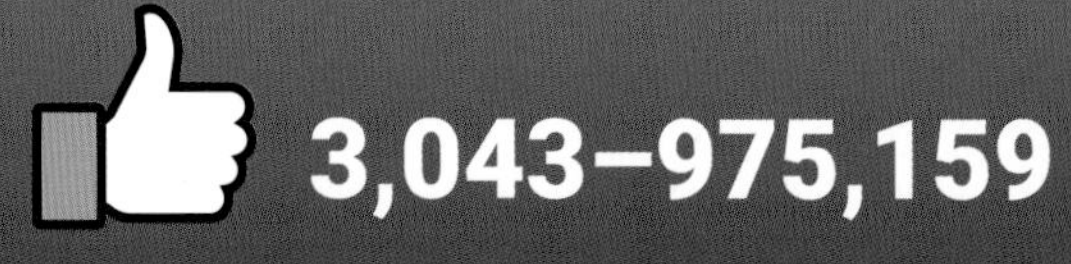

From the moment the social media bug bit me, amazing things began to happen to my blog! All of a sudden, it wasn't just my family who checked it out to see me pushing Elliot's face into cakes, crumbles, trifles and flans. The whole of Bridgend signed up to check out what pudding I'd stick in front of him whenever he finished his main course! My brother fell for it every time. And I became the talk of the town.

Then prank fans from beyond Bridgend began to visit my site, and I was forced to think hard about where this was heading.

'We don't even know what people *look* like north of the motorway,' Elliot warned me one day, once he'd wiped the treacle tart from his face and calmed down a bit. 'They might be dangerous, Ben. I DON'T WANT TO DIE!'

'But everybody loves you, bro,' I said before showing him how I planned to take things to the next level.

'What's happening?' he asked as I held up my new phone and framed him on the screen. 'Why you recording?'

'Just relax,' I said casually, and began to follow him through the house. 'It's nothing to worry about.'

Elliot turned and scowled at me. I'd just steered him towards the kitchen, where he found a plate of sliced carrots waiting for him. Now, if there's one thing my bro can't resist, it's carrots with a good splurge of ketchup. Which is why I'd left the bottle beside the plate.

'Is this for me?' he asked.

'I thought it might make you feel better,' I told him.

Elliot glanced over his shoulder, but I knew he couldn't resist. As he found himself a fork, I settled on a stool on the far side of the kitchen and quietly continued to film.

'Ben, stop recording!' he said and began to shake the ketchup bottle. 'It's just weird is what it is. Weird and creepy!'

'Just calm down,' I said, because I knew that would wind him up even more. 'And maybe check that ketchup lid is on tight. You might make a mess with all that shaking.'

As I hoped, Elliot just shook the bottle even harder.

'I can manage a ketchup bottle, Ben. Next you'll be suggesting I need a pair of goggles or something.'

'That's a good idea,' I muttered.

'What?'

'Oh, nothing,' I said innocently, and even put my phone down beside me.

With his eyes fixed on me, Elliot gave the bottle one final shake. Then he grasped the lid, ready to lash some sauce all over his favourite sliced veg.

'You don't know what you're missing out on here!' he said, only to grunt because the lid seemed to be a bit stuck. 'Carrots and ketchup is a classic.'

Watching Elliot struggle with the lid, I reached for my phone once more. The bite on my hand had healed, I felt fine, and my social media senses told me that what was about to happen would make a brilliant debut on my new platform. My food blog was creaking under the traffic and now I was planning to take things into a whole new universe of fun.

Elliot had been pied more times than he'd had hot dinners, so I'd decided to take things to a whole new level and added baking soda to the ketchup. Why? Because when that stuff is shaken up it causes a chemical reaction that turns an average tomato sauce into a ketchup cannon. With the pressure building up inside the bottle, it was no wonder Elliot had trouble undoing the lid.

'Everything all right?' I asked.

'Everything is *fine*!' he grunted. 'Here it comes ... '

As he began to twist the lid free, I felt sure this was just the first of many pranks that would go far beyond our little Welsh town. I wanted *everyone* to laugh along with us. As far as I was concerned, there could only be one way for me to share my brother's misery with the world: Facebook.

iPad
9:41 AM
91%
Timeline
Ben Phillips
30 March at 16:10 •
Ketchup Bomb!
Like
Comment
Share
Wilf W, Dai Burdock, Bluebell & 112K others
Top comments
19,000 shares
Grace Rendell This is so funny! poor Elliot!
Like • Reply • 245 • 30 March at 16:19 • Edited
6 Replies
Mrs Phillips LOL! Is this what you get up to when your father and I are out?
Like • Reply • 799 • 30 March at 17:03
45 Replies
Nana Get out of my kitchen!
Like • Reply • 351 • 30 March at 17:03
14 Replies

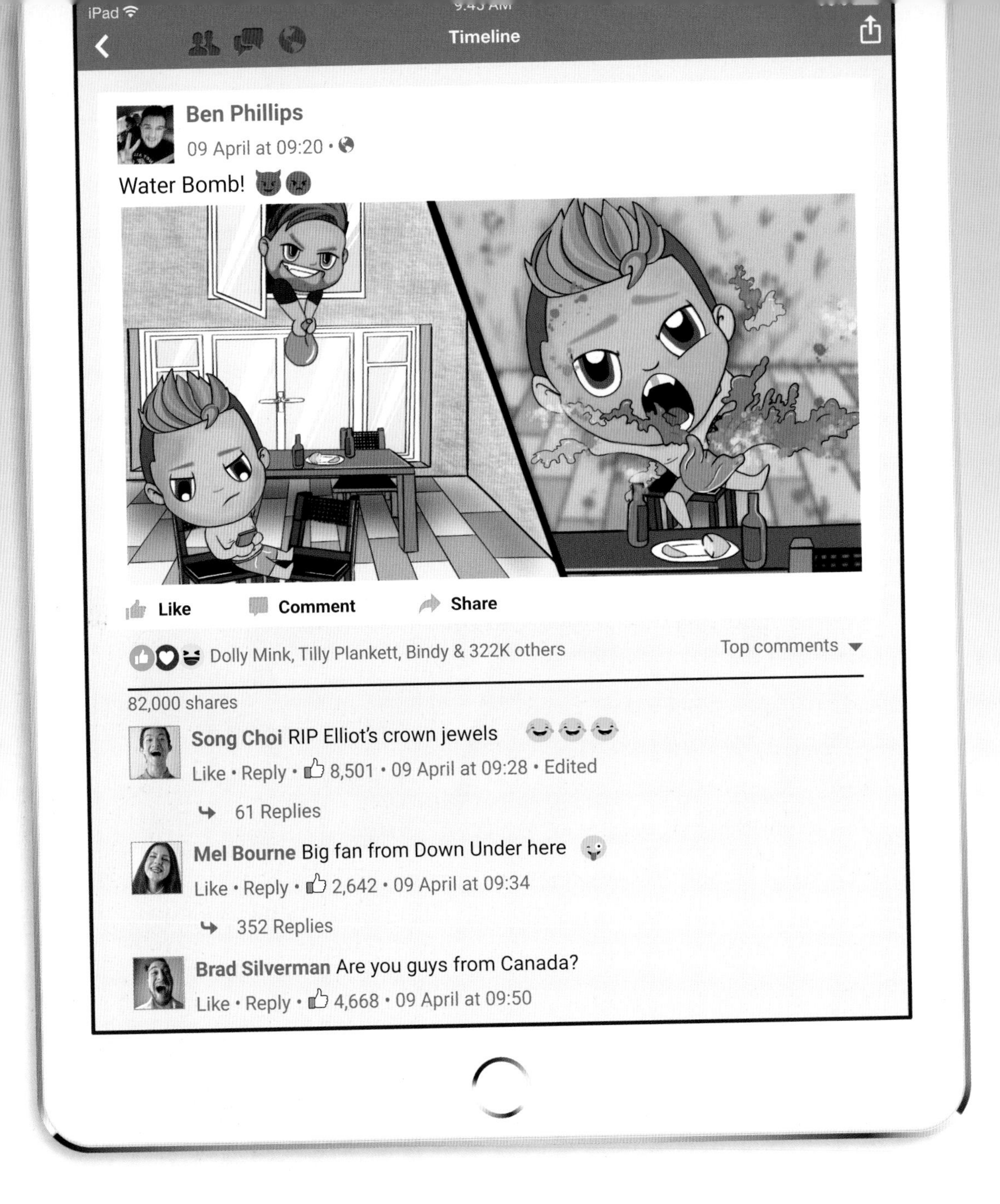

Within weeks, as my pranks reached out around the planet, I became a charity of smiles. Laughter filled the air, and not just from downstairs whenever Mum and Dad were checking out my latest video. Moments after each upload, I could hear chuckling in the street and over the fields. As the comments rolled in from left, right and centre, I really felt like I had put my super social media powers to good use.

'The happiness is spreading, see,' I pointed out to Elliot one day as we made our way across town. 'And it's all down to you, bro.'

Everyone from babies in prams to old ladies passed us by with a cheery nod and a wave. Even the sun seemed to sparkle, and Bridgend had never known anything like that. In fact, only one person failed to share the joy. Frankly, every new prank just left him with a face like a bulldog that had just licked wee from a stinging nettle.

'There's NOTHING funny about this!' screamed Elliot as he struggled to keep up with me one day. No surprise, though, as he had the toilet seat glued to his bum. 'It's stuck tight, Ben! And Dad is going to kill me when he sees that I've had to unscrew it from the bog!'

'How else are we going to get to the hospital?' I asked, trying hard not to roll around laughing. 'Just keep hold of that sheet around your waist, Elliot.'

'It looks like I'm wearing a *skirt*!' he growled.

'Yeah, but you don't want to drop it and risk being done for flashing.'

Elliot grumbled all the way to A&E. Two hours later he left with the loo seat under his arm and a sore backside. Me? I laughed all the way home. At the same time, despite my mission to spread joy across the internet, I felt a little bit sorry for my brother ... and that's why I arranged for Georgina to come into his love life.

ELLIOT'S DIARY

I'm going on a date! At last, after all the grief from my idiot brother, it looks like my luck is about to change.

A local girl's texted to ask if we could meet. I know, right? She follows Ben's Facebook page, and says she's a big fan of my work. I was going to point out that I'm just an innocent victim, and that people like her who watch Ben's work only encourage him. But then she sent a picture of herself. It made me feel funny all over. Without thinking, I replied with an emoji of a big thumbs-up and a face with the tongue hanging out. I know how to keep it classy, see.

I'd never met Georgina before. Ben always says how Farmer Chris' sister is in a class of her own, but I didn't realise just what that meant until I got to see what she looked like. The picture was a little shadowy, and very dark around her chin, but she has amazing eyes. Very inviting. I was in love before we'd even met!

'How much gel have you used?' asked Ben as I left the house that evening. I actually had to duck under the doorframe so I didn't damage it with my hair where it stuck up at the front. 'And what's that funny smell?'

'Aftershave,' I said. 'Pure Bull.'

'It certainly is!' Ben pulled a face. And then he flashed a grin at me. 'So where are you meeting Georgina?'

I was out on the path when he said this. It was enough to stop me in my tracks, turn and frown at him.

'How did you know I'm seeing Georgina?'

Ben seemed lost for words for a second, only to look like he'd suddenly remembered something.

Elliot STINKS of this

'Everyone is talking about it, bro,' he beamed. 'You're dating the dream!'

Part of me knew he was up to something, but the other part was well aware that if I hung around and questioned him I'd be late. Ben must've noticed me check my watch, because he told me to get going.

'Believe me,' he said. 'You don't want to show up looking sweaty and out of breath. Georgina is a hot babe with high standards.'

'I know,' I replied, and tucked my shirt in for the tenth time that evening. 'That's probably why she wants to meet me.'

Twenty minutes later, at the restaurant, I found myself waiting nervously for my date. I kept one eye on the door. I was looking for a 'total goddess', as she had put it in her text, but nobody came in matching that description. So I stared at the menu, wondering how long she would be. A moment later, a shadow fell across my table. At the same time, it felt like the temperature in the room had dropped by a couple of degrees.

'Elliot?'

I looked up with a start. The woman who had approached my table wore a hat with a broad, floppy brim, and spoke with a high-pitched voice that still managed to sound like grit in a cement mixer.

'Who's asking?' I said nervously.

She grinned and dropped into the seat opposite.

'Georgina.' The vision before me extended a beefy hand and squeezed mine so tight I thought she might crush bones. 'Are you ready for a night to remember, handsome?'

If her voice and handshake were enough to silence me, her big beard and sideburns almost stopped my heart from beating. With my jaw dropped in shock, all I could do was stare. At the same time, as I realised I had been set up, a small voice in my head began to shout: Ben . . . ? Ben! BEN!

'Will you excuse me for a moment?' I said. 'I just need to visit the little boys' room.'

'Surely not that little?' Georgina took a swig from her drink, eyeing me hungrily as I squeezed out from behind the table. 'Don't forget me,' she added with a wink.

There was no danger of that. As I scrambled out of the window in the bogs and then took off across the fields for home, I worried that what had just happened would haunt me to the grave. I ran as if I expected a pack of dogs to come chasing after me at any moment. I was so freaked out that Georgina might follow that I didn't even stop to check for traffic when it came to crossing the main road. I just rushed out into the path of a car, and then froze in horror at the screech of brakes. All I could do was gulp as the open-top car halted just inches from me, and then remember to breathe as I came face-to-face with an angel behind the wheel.

'Are you OK?' asked the vision that rushed from her seat to check on me. Her name was Amelia, as she told me later. She was taller than me by a couple of inches, with blonde hair, freckled cheeks and no sign of the kind of facial hair you'd expect to find on a lumberjack.

'I'll be fine,' I said as our eyes met. 'In fact, I'm beginning to think I've never felt better!'

Elliot, if you wanna
get with the ladies
you need sex ap-peel

GETTING TO GRIPS WITH GEORGINA

She's the girl who can kill a conversation just by walking into the room. Farmer Chris' sister is one of a kind, all right. Mostly that's down to the beard, and the squeaky voice that always sounds like it's about to drop into something surprisingly . . . well, manly.

No offence, but what's with the beard?

What beard? How rude! Ben, my skin is as smooth as your brother's chat-up lines. I just wish he'd use some of those lines on me.

We all know you like him, Georgina. I just don't think he's that hot on you.

You don't know him like I do, Ben. He wants me. I can tell. He wants me bad. If you'd just stop pranking him for 10 seconds,

then perhaps he'd have a chance to realise. Instead, he's too busy gulping down milk because you've slipped chilli into his meal, or walking funny because you've messed with his chips and given him the runs.

You're alone with Elliot and he's unable to get away. What would you do?

What wouldn't I do? Ben, there are some things a lady can't say without embarrassing herself. Let's just say I'd leave him begging for mercy.

What do you like best about Elliot?

His black front tooth. It marks him out as special, and I mean that in a good way.

If you could change one thing about him, what would it be?

His brother.

Elliot's Dating Profile

When he's not sleeping, eating, or ranting at me, my bro likes to date. For ages, I've been wondering what he's got to offer that makes him so popular. It's really weird. So, I did some digging around the last time he went out and I found the secret behind his success. No, he isn't bribing girls. It's all down to his dating profile. You really should be wary of strangers . . .

Elliot, 19

0 2

Girls, I'm warning you now. Elliot doesn't look like the picture of James Bond he uses. He doesn't nurse poorly kittens to health for a living and there's no way on earth you'll find a pound in his pocket, let alone the million he claims to have in the bank. I decided to set the record straight so his profile no longer lies . . .

●●●○○ Bro-mobile 4:12 PM

hot__boyos.com

Welsh lads looking for laydeez

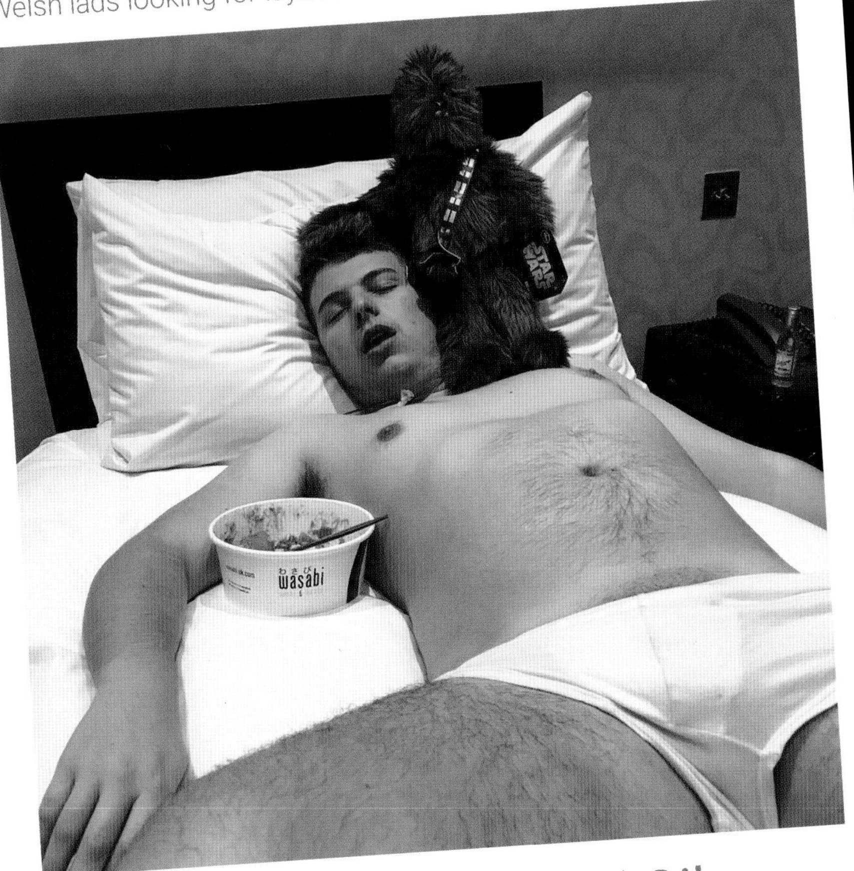

Elliot 'Single as a Pringle' Giles

2 miles away Active 8 hours ago

Age: Yes

Education: Sexytime

Previous dating experience: Nope

Interested in: Anyone but Georgina. Must have a mother and a face. Actually, I'm not bothered about the last bit.

Body Type: Chafed (mostly around the nipples because I had to sandpaper down the special glue that Ben used to stick on the rings).

Your life in a sentence? Full of shitty pranks.

Your perfect date? One that doesn't end in TOTAL DISASTER thanks to my brother.

THE SCIENCE OF PRANKING

Illustrations by Nathan Balsom

It looks easy, but there's some carefully balanced equations behind the misery I inflict on my bro...

BEN'S BEAUTY TIPS

Looks aren't everything. Ask Elliot! Even so, my bro takes great care of his appearance – especially before a date. Honestly, he can spend an hour locked away in the bathroom. I've no idea what he gets up to in there. Actually, I do – having rigged up a secret camera – but I'll spare you the details. Save to say that when he steps out reeking of aftershave, and with so much gel in his hair that a downpour of rain could cause an oil slick, my bro thinks he looks hot to trot.

Which is why I always try to help him out when he takes a nap, and make him more presentable. When it comes to bringing out Elliot's true beauty, I'm all about creating an impression that nobody can forget! Here's how:

Guy Liner

They say the eyes are like a shop window to the soul. In Elliot's case, once I've finished my work, his look more like a horror show. It's all down to careful pencil shading underneath each eye. I say pencil. I mean solid black permanent marker, drawn on while my bro is snoring.

Luscious Lashes

Speaking as an experienced make-up artist (working with just one person), it's really important to do a thorough job here. You can pick up fake lashes from most beauty stores, but I recommend a fat dollop of special glue to make sure they stay in place. The last time Elliot went out with no clue that he was wearing a pair, his poor date found herself fanned with a blast of air every time he blinked. Useful in summer time!

Ballsy Blusher

A little colour in the cheeks is healthy. A lot of colour in the cheeks is, well . . . what you expect to see on a clown at the circus. Use a red permanent marker or, if you want to go large, reach for a pot of doorstep paint. You can be confident it'll last the length of a date. And the week that follows.

Lingering Lippie

There's a reason why girls are always going off to check their lipstick in the mirror. Because it always comes off! So don't bother putting on standard lipstick. In fact, you can save money by using the same red permanent marker that you've

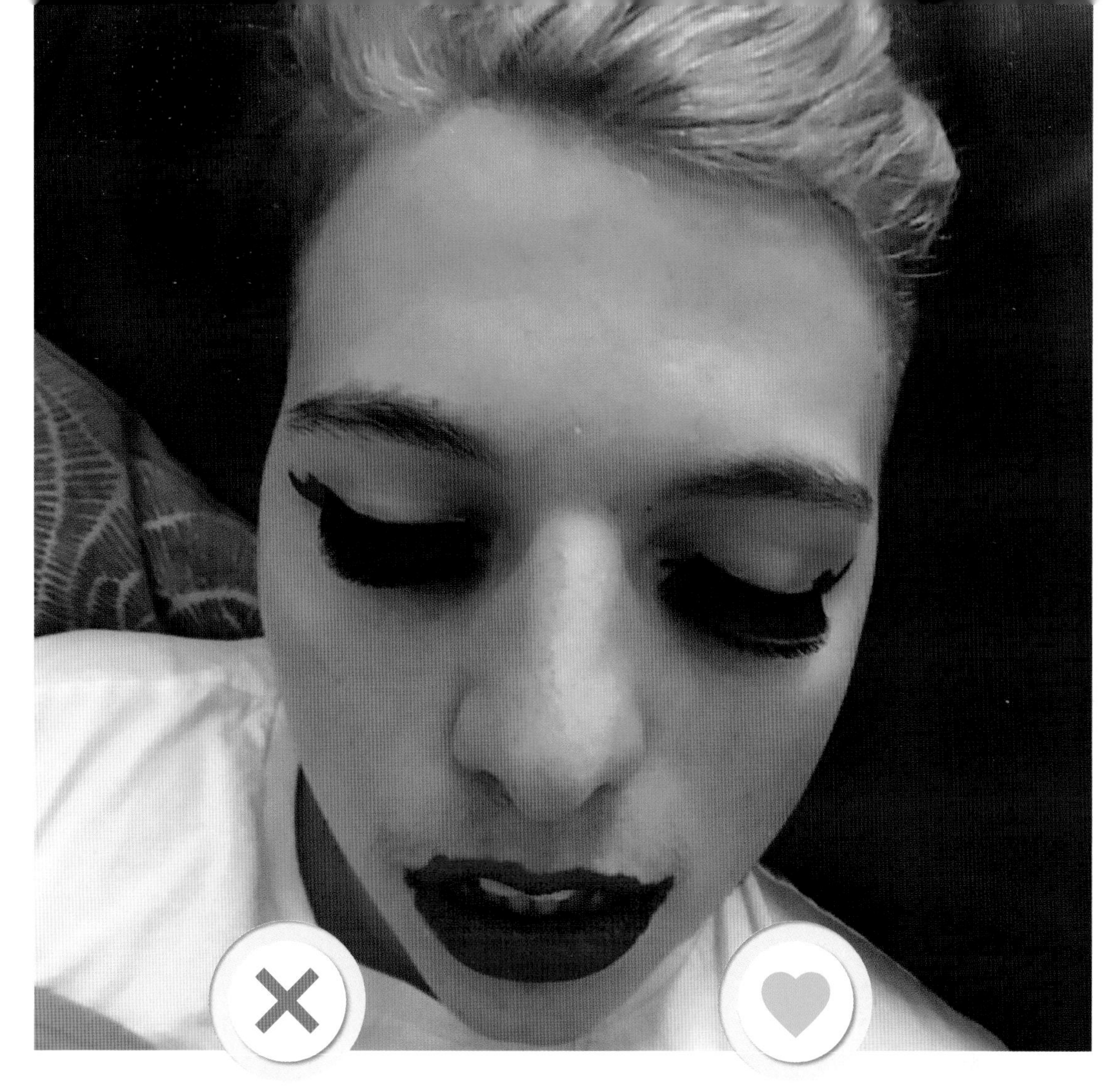

just used to rosy up the cheeks. If I really want people to pay attention when Elliot walks into the room, I finish the job with a little special glue and then sprinkle on the glitter. Magic.

Facial Haircare

A good shave can show that a man knows how to take care of himself. It's a shame we can't say the same thing for Georgina, but Elliot makes the effort. As his bro, it's only right that I help him out here. I'm just not so keen on leaving him with skin as smooth as a baby's bum. At times I like to take things in the other direction by manning him up as much as I can. The special glue comes in handy here, whether you're going for a subtle comedy moustache or sprinkling on chopped pubes for a stubble effect with a twist. Whatever, when it comes to male grooming I always aim to make sure my bro turns heads (and stomachs).

Chapter 7

I couldn't believe it! How could my brother head out on a doomed date and return looking like the cat that got all the cream and more?

After coming face to hairy face with Georgina, I'd expected Elliot to trudge home feeling deeply sorry for himself. I was waiting for him, in fact, with my phone camera set to record as I questioned him about what had happened when everyone's favourite bearded lady sat down at his table.

Instead, a sports car screeched to a halt outside our house with a smoking-hot blonde behind the wheel. At first I thought she must be lost. Then I saw my brother in the passenger seat. She leaned across to kiss him on the cheek, and I dropped my phone in surprise.

'Amelia refused to let me walk home!' Elliot explained later, once I'd got over the shock. 'And she'll be coming back tomorrow evening to pick me up,' he mumbled, looking at the floor. 'For our date.'

'But ... but you can't!' I protested. 'She's way too good for you.'

Elliot shrugged but didn't argue. I had been planning on shoving my camera into his face to capture his anger at being set up with Farmer Chris' seriously weird sister. Instead, I just stood there wondering if *I* had been made to look a fool here.

There and then, I decided that I had to stage a prank or two that would put Elliot in his place. My newfound social media powers were practically calling out for me to grab the special glue and put it to good use. I considered glueing him everywhere from the bath to his bed and even the ceiling, but it wasn't enough. This time, I had to push Elliot so far into the anger zone that Amelia would see through his simple charm and run a mile.

What I hadn't considered was the power of love. Elliot was smitten. Completely head over heels. As a result, no matter how hard I tried to prank him throughout the next day, there was no way my bro was going to lose his cool. Even when the joke was at his expense, he never failed to see the funny side …

AARGH! He's too happy!

Honestly, I lost count of the number of pranks I pulled that day. Thanks to my social media powers, every single one should've left my brother fuming while ramping up my followers. Sure enough, the view count climbed, but Elliot just laughed off my efforts before drifting away to daydream about his date.

With an hour to go before Amelia showed up to sweep him off his feet, my brother did what comes naturally. Having showered, dressed up and gelled his hair, he settled on the sofa to wait for the doorbell and promptly fell asleep … I mean, come on! It would've been wrong to just leave him alone. I only had to hear Elliot snoring once and my nose for a hit clip demanded that I take advantage of the situation. Straight away, I headed for the kitchen drawer, and returned with a marker pen in hand.

ELLIOT'S DIARY

I didn't hear the doorbell ring. I'd been enjoying a lovely dream that involved lying on a desert island made from jelly, only for Georgina to jump on me and shake me awake. I opened my eyes with a gasp, to discover it wasn't Farmer Chris' bearded sister in my face, but my brother, Ben.

'Wake up,' he urged, and shook me again by the shoulders. 'Amelia is here!'

'What?' I sat bolt upright so smartly that Ben fell back on the floor. 'Oh, sorry, Bro!'

'Never mind that,' he said, and scrambled to his feet. 'She's waiting for you outside!'

'I just need to check I look all right,' I said, turning for the mirror.

'There's no time!' Ben spun me around full circle and shoved me towards the door. 'Now go make an impression on her that she'll never forget!'

I had to pinch myself when I saw Amelia waiting for me. There she was behind the wheel of her sports car, with the engine throbbing away, and she couldn't take her eyes off me! Seriously, I climbed into the passenger seat and she just carried on staring with a smile that seemed to grow.

'Everything OK?' I asked.

Amelia responded with a blink as if a magic spell had just come to an end.

'Elliot, you're so sweet,' she said, kissing me on the cheek. 'I love your honesty.'

At first I thought perhaps she meant I'd left my flies undone. A quick glance told me I was all good on that front, and so I took the compliment. I even sneaked a thumbs-up at Ben, who was at the front-room window. Then I realised he was filming me, and so I left him with a scowl.

'Let's go,' I said to Amelia. 'It's just you and me, chicken!'

That evening, after a cosy dinner for two, we drove up into the hills overlooking Bridgend and looked down upon the town. Lights twinkled far and wide beneath a clear and starry sky. We've could've been in Los Angeles, wherever that is. All right, maybe I was pushing my luck when I said that out loud, but Amelia seemed impressed. She snuggled into my shoulder and didn't seem at all worried about what might lie behind us.

Being so close to the town border, I kept glancing over my shoulder into the darkness and the mist. As a Bridgend boy, I had no idea what lurked beyond the hills. Monsters, most likely. The last thing I wanted was for this perfect date to be ruined by one. Still, I realised I could keep an eye out without turning round by flipping down the sun visor and using the mirror. I just hadn't planned on catching sight of my own reflection as I did so, or lurching forward in horror as I took in the slogan scrawled across my forehead.

I LOVE GERBILS!

'What's the matter, babes?' asked Amelia, stirring as I squeaked in horror.

'What's the matter?' I jabbed a finger at my forehead. 'Look at what he's done!'

She glanced up and tried hard not to giggle. 'I think it's sweet. I love a man who's unafraid to show his commitment to defenceless animals.'

'Argggghhh!!!'

With the red mist descending, and feeling totally ashamed, I jumped out of the car and started to run. I didn't care about the monsters any more. If any swooped overhead they'd be sorry. That's how mad I was at my brother.

'Ben!' I yelled as I barrelled towards town. I probably woke up the neighbourhood just then, but I didn't care. Something terrible was about to happen all right … just as soon as I got my hands on my brother. 'Ben! BEN!!'

I LOVE GERBILS!

16 messages from:
ANNA #7, HOT HEIDI,
NAUGHTYNIA...

1 message from:
GEORGINA

HOW (NOT) TO CHAT TO GIRLS

I slipped an iPhone inside Elliot's pocket every day for a week. The results are pretty shocking, and the reason why he's single. If you want to know how* not *to impress a girl, here's how to do it, Sorry Bro-***style!***

Excuse me, I'm lost. Could you show me the way to the kebab house? No, wait. That's wrong. I mean your heart? Your heart!

You remind me of a parking ticket . . . because you're all yellow. Think it must be the fake tan. What's your secret?

Is your name Georgina? If it isn't, you've pulled.

If you were the only girl in the world, and I was the only boy . . . yeah, I know. Beggars can't be choosers but . . . oh, come on! Don't go!

Your place or mine? Actually, can we make it yours? My brother's at home.

What's a nice girl like you doing in a dump like Bridgend? I mean, c'mon. All the pretty ones left ages ago. No offence, like. It's just . . . what?

Get your coat, love. Nah, you haven't pulled. This is Bridgend. It'll get nicked if you leave it lying around.

Definition of true love? You can watch telly at my house and I'll let you hold the remote.

You're as beautiful as one of those farts I can drop silently at the table before blaming Ben.

As soon as I saw you my heart started hammering like my fists on the inside of the coffin when my brother buried me alive.

Can I just check something? When you said I'm special, did you mean that in a romantic way, or . . . oh, right. Well, you're wrong about me. YOU DON'T KNOW WHAT YOU'RE MISSING!

I love kissing with my eyes open, I do. But that doesn't make me a serial killer. You don't have to worry on that score.

I'd walk 100 miles just to be with you. Unless you're on a bus route, obviously.

Me, you, fish and chips for two. Sorted, right? RIGHT?

So, you went to school with Nana? Really? You don't look that old . . . I know. Smooth, right? So, does that mean you'll go out with me? Result! Um . . . can we make it our secret?

Elliot's Love Poetry

If Elliot has one weak spot, it's his heart. Not in a medical way. He's just a bit soppy when it comes to girls. I'd like to say he's had more dates than hot dinners, but we only need to check out his jelly belly to know that isn't true. Still, he's a romantic at heart, and I've got the bad poetry to prove it.

They're private, of course. Elliot keeps them hidden under his bed. So, once I've shared them with the world I'll put them back and we can make it our secret, right?

ME, YOU & A BURGER FOR TWO

Last week we sat in the park
under moonlight
Munching on chips with cheese.
I had a burger, too. You didn't want
one. Then you changed
Your pretty mind.
So, we took turns eating mine, until
there was nothing left
And that's when our lips met
for a kiss
I hope I didn't reek of gherkins.

FRIDAY NIGHT AT THE BOWLING

Strike out!
That's what you do to my heart
Every time I see you.
You handle those big balls like a pro.
And bowl me over every time
You thrashed me, in fact.
Which was embarrassing
And left without a kiss.
Even though I paid for the tickets.
Oh, well.

SORRY ABOUT MY BROTHER

I can explain
Everything.
It was my brother, see?
He thinks it's funny to prank me.
So, that meal we shared didn't work
out well
As I think you can probably tell.
I can only blame the fart powder
Which couldn't have made things louder
I just wanted to say that I'm sorry,
all right?
Hope you stopped retching.
I love you.
Good night.

ELLIOT

Height: Quite short (and fat)
Strength: −2
Speed: 13 (when he's angry, 3 when he's not)
Occupation: Benefits
Battles won: 0
Likes: Some weird stuff, having a beer and then four more beers
Dislikes: BEN!
Secret weapon: Rage
Special move: Wetting himself

BEN

Height:	Very tall
Strength:	10
Speed:	9
Occupation:	Prankster
Battles won:	2,369
Likes:	Nerf guns, having a laugh and pranking Elliot
Dislikes:	Elliot's turtleneck
Secret weapon:	iPhone
Special move:	Gluing Elliot to things

CRAP CRIBS

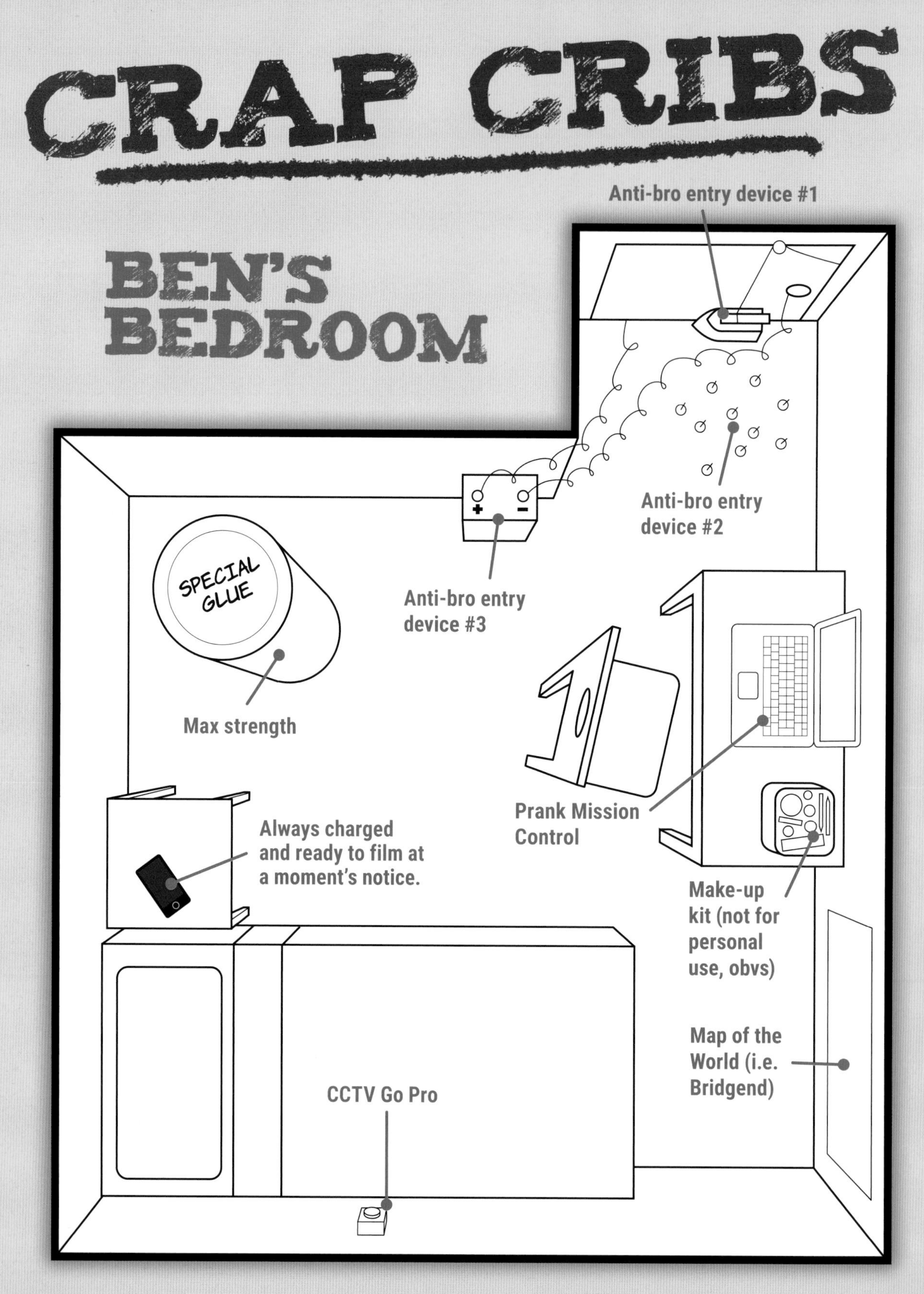

BEN'S BEDROOM
Anti-bro entry device #1
Anti-bro entry device #2
Anti-bro entry device #3
SPECIAL GLUE
Max strength
Prank Mission Control
Always charged and ready to film at a moment's notice.
Make-up kit (not for personal use, obvs)
Map of the World (i.e. Bridgend)
CCTV Go Pro

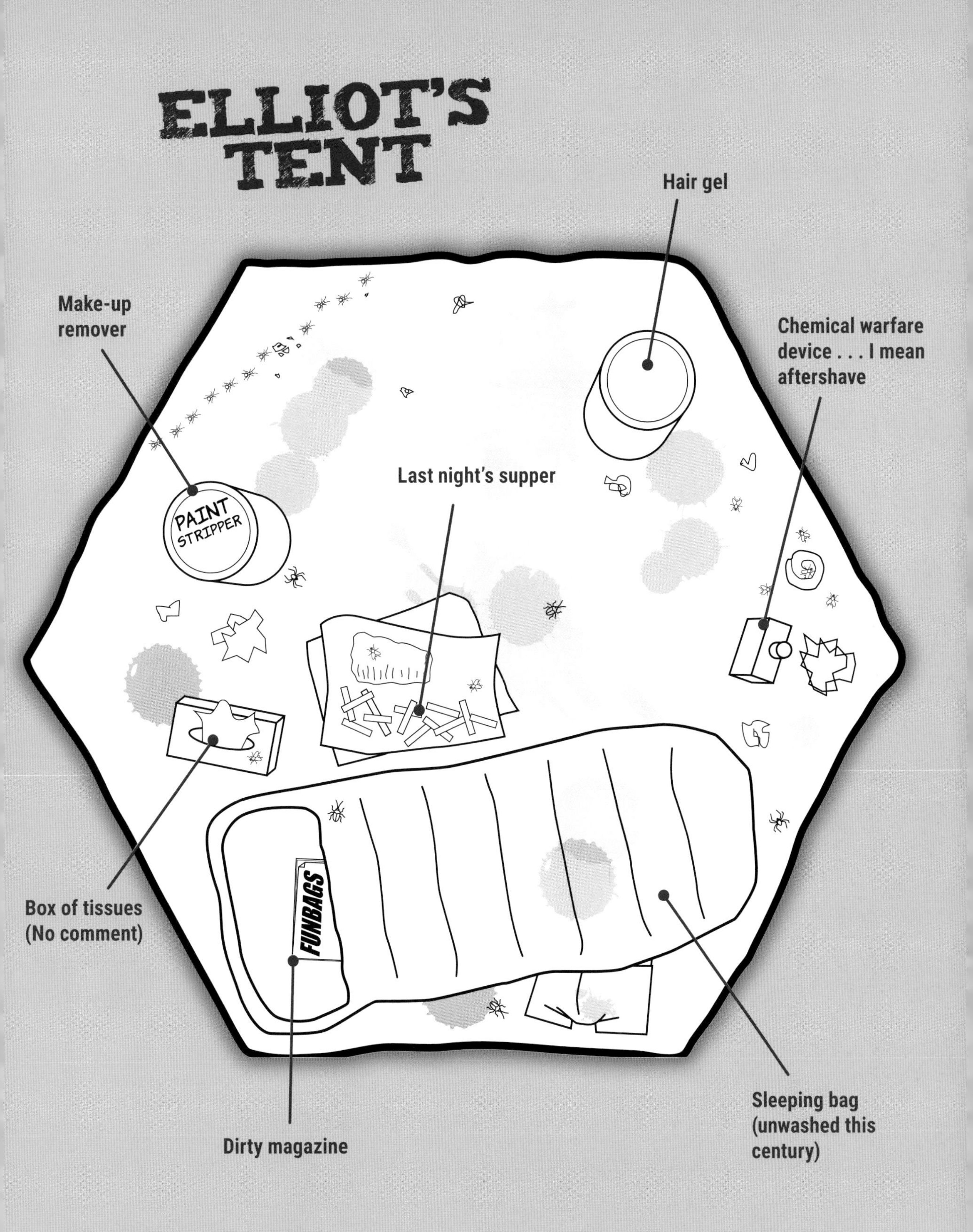
ELLIOT'S TENT
Hair gel
Make-up remover
Chemical warfare device . . . I mean aftershave
Last night's supper
PAINT STRIPPER
FUNBAGS
Box of tissues (No comment)
Dirty magazine
Sleeping bag (unwashed this century)

Chapter 8

1.4 million–4.7 million

A week went by before Elliot calmed down. It took the same length of time for the marker pen to come off. By then, if anyone dared mention gerbils, my brother just *exploded*.

'Why?' he ranted at me one time on catching sight of his reflection and the fading statement on his forehead. 'You ruined the perfect date, Ben! I don't suppose Amelia will ever talk to me again!'

'You've always got me, bro,' I chuckled. 'Plus close to two million fans!'

This was enough to silence him. I was checking out my Facebook page at the time. Earlier, I uploaded my recording of the moment Elliot had stormed home from the hills and raged at me. I hadn't been able to stop laughing, which only cranked up his anger by another notch. On the upside, the clip had brought in hundreds of thousands of hits. My social media powers were *singing*, and Elliot was the star attraction! My brother peered at the screen and then faced me.

'This,' he said, 'has got to stop.'

Who DOESN'T love them?

'But you're famous!'

'I don't care, Ben. You're making me look stupid!'

'But you *are* stupid,' I reasoned, 'when you lose your cool.'

My brother narrowed his eyes at me. 'Just give me a break, OK?'

I raised my hands in submission.

'Whatever you say,' I replied, and quietly crossed my fingers.

To be fair, I made sure I covered my tracks for the prank that followed. I had figured out that soaking a toilet roll in water, scrunching it up and then rolling it between my hands could create a convincing, dark brown poo. For the next few days, Elliot would find these little presents everywhere he turned, from his bed to the bathroom and the back seat of the car. Weirdly, he cleared up each one without complaint, as if perhaps he thought he'd made the mess himself and didn't want anyone to find out. I quietly filmed each one, of course, and put together a compilation that went viral overnight.

My social media powers were growing bigger than I ever imagined. I only had to upload a video to cause such a surge in traffic that servers around the world came close to melting down. I craved the hits, fans and followers, and found myself constantly seeking an opportunity to prank Elliot.

So when his phone rang towards the end of that week, and Amelia's face flashed up on to the screen, I sensed an opportunity shaping up that I just couldn't ignore.

'She wants to see me again!' my brother crowed afterwards. 'First she said how sorry she was for not telling me what you'd written on my face. Then she asked if I'd like to go the funfair with her.'

The moment he told me this, I began to plan for a prank that would bring Elliot to the attention of the world and beyond! Watching my brother swagger away, I knew exactly what I had to do.

The clown wig came from the party shop. With wild red curls frizzing out all over the place, I knew it would be perfect for what I had in store. OK, so it was a little bit big for my brother's head, but that was easily fixed with a generous splurge of special glue before I popped it on as he lay dozing in front of the telly.

'Is that the best you can do?' he asked on waking. Elliot stood in front of the mirror, tugging in vain at the headpiece.

'The finest,' I told him, as he grabbed it with both hands and pulled. 'I would help but I'm filming, see,' I added, and took a step back with my phone in hand in case he kicked off.

But Elliot didn't lose his cool. Instead, he gave up trying to get rid of the glued-on wig and simply glared at me.

'You're forgetting that Amelia likes me for who I am, not how I look,' he said gruffly. 'She's not shallow like you, Ben.'

And with that, Elliot left me to get ready for his date. He was steaming about the wig, of course, but convinced that he had found a girl who understood my pranks were beyond his control. It left me with no choice. If my brother was set to spend an evening with Amelia at the funfair, I would have to stage the mother of all pranks – something neither of them could ignore.

Thanks to a bite from a social media bug, I had powers to amuse that couldn't be stopped. Elliot had no choice in the matter, I thought to myself as I slipped out of the house to put my plan in place. I felt driven just then, as if I was answering an urge to upset my brother and make millions of people laugh. In a way, my gift had come close to a curse that went way beyond my control.

ONE HOUR LATER . . .
VRROOOOMM
RRMMMMBLL
OH, ELLIOT, THIS IS THE BEST DATE EVER!
IS IT? WOW!
WHAT SHALL WE GO ON NEXT, BABE?
HMMM...
TICKET BOOTH
THAT!

PERFECT! I LOVE A THRILL, I DO.
I JUST KNEW THEY'D HEAD FOR THE ROLLERCOASTER, AND I WAS READY.
EARLIER, I'D PERSUADED THE GUY TO HELP ME OUT.
TICKET BOOTH
£1
ALL I HAD TO DO WAS GIVE HIM THE SIGNAL . . .
THAT'S A POUND FOR EACH OF YOU, MISS . . .
. . . AND CALL IN MY SPECIAL GUEST.

ELLIOT'S DIARY

As the train began the slow climb towards the top of the rollercoaster, I'd never felt so happy in my life. There I was beside the girl of my dreams, with the wind blowing wild through my clown wig. OK, I could've done without the wig, but Amelia didn't care and nor did I. We were together, and there was nothing Ben could do to tear us apart. Or so I thought.

'Are you ready?' she asked as we neared the crest.

I have to say Amelia sounded a bit shrill just then. I put it down to nerves.

'Actually,' I said, grasping the guard rail, 'I'm a bit scared.'

'I'll protect you,' she chuckled, in a voice that sounded horribly deep all of a sudden, and then screeched as the train plunged down the other side.

By the time we'd been round once, I was worried I might lose my lunch. The last thing I wanted to do was be sick all over my date, so I squeezed my eyes tight shut and thought of my special place. As the train climbed once again, I pictured myself back in the woodland glade. All my furry friends were there, as well as the earthworms, and that helped me to feel safe. Then the train whooshed into a drop, and in a heartbeat I was screaming my lungs out.

'Isn't this fun!' I heard Amelia yell in my ear as we turned and twisted at high speed.

'No!' I whimpered, and just buried my face into her shoulder. I didn't care any more. This rollercoaster ride was a huge mistake.

A moment later, when I caught sight of a familiar figure watching us, I knew for sure that I should've steered well clear.

'Ben?'

It was just a glimpse. A face in the crowd as we whizzed around, but that was enough to send a chill down my spine. It didn't help that I had spotted him filming.

'What's wrong?' I heard Amelia ask, her voice as piercing as the shrieks from the passengers behind us.

'Oh, nothing,' I said, and braced myself for the section where the track went into a loop-the-loop. Then, as we turned upside down, all the power just drained out of the ride as if someone had hit a switch. I looked down and saw the guy in the booth leaning back from a lever. At the same time, the lights went off, as did the music, and the crowds below gasped as one. In shock I looked at Amelia. It was weird to see her upside down, with her hair hanging loose.

It was even more of a shock to see that she'd grown a beard all of a sudden.

'Georgina?' I said in a whisper.

'Be strong, big boy,' she said, before planting a kiss on my lips. 'I'll protect you!'

'Nooooo!'

CReam

I tried to escape from Georgina's clutches, but we were buckled in tight. In the struggle, the bag at her feet popped open. First a lipstick fell out and dropped to the ground, followed by a can of whipped cream and a pair of fluffy handcuffs.

'Oh, never mind,' she growled as I fought off her wandering hands. 'We can still have some fun while we wait to be rescued.'

'Ben!' I cried out, well aware that everyone was watching a fright clown tussling with a bearded lady, trapped upside down on a rollercoaster. It explained why my brother wasn't the only one with his camera. 'What have you done with Amelia, Ben? Ben! BEN!'

As a firm hand hauled my face round for another kiss, I glanced down and saw him once more. There he was, creased up with laughter, which told me exactly what had happened here.

'This clip is guaranteed to break the internet!' Ben called up.

'Sorry, Bro!'

So, today I mixed Elliot's coco pops
with Norman's littertray. Sorry, Bro!

KITTY Q&A: NORMAN THE CAT

Cats are clean, hygienic creatures. For one thing, they won't poop on their own territory. Take our next-door neighbour's moggy, Norman. Whenever he needs to take a dump, he'll jump the fence and find somewhere quiet, comfortable and in tune with his animal nature . . . Sorry, Bro!

Tell me about your life

Firstly, Ben, can I just say what a surprise and a joy it is to discover that you can talk to the animals. I had no idea! You're like Bridgend's very own Dr Dolittle. That's a step up from Elliot – known by the local pet community as Dr DoNothing. Now, to answer your question: life is good! Just as long as I keep one eye open for your garden-fork-throwing brother, I'm happy. What's wrong with him, though? That boy has serious anger issues.

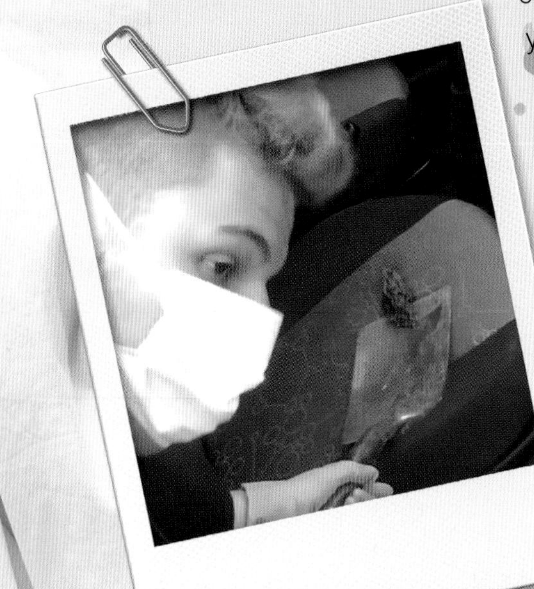

Let's just say he's not a cat-poo person.

Dude, I'm just doing what comes naturally. I'm simply finding a place to relieve myself that causes least offence. I can't understand why

Elliot gets so upset. From his car to his clothing, everything that wild boy owns smells kind of earthy in the first place. For basic creatures like me, it's an open invitation to leave a special gift.

You do realise it upsets him, don't you?

Yeah, but there's a difference between being upset and completely murderous! When Elliot's mad at me, I can hear him yelling my name from the end of the street. As for the number of times he's tried to assassinate me, I'm lucky to be alive! That boy needs to calm down. Otherwise, there could be consequences . . .

Consequences? Like what?

My owner, Marge, might seem like a sweet old lady, but between us she runs the biggest catmint smuggling network in the UK. She knows people, Ben. Bad people. You get me? Perhaps it's time you had a word with your brother, and reminded him of his manners.

THE 'PRANK ELLIOT'

People are always coming up with crazy (some say sick) ideas for pranks I can play on my bro. So I set up a suggestion box and here is the very best of the bunch – or the worst, depending on how you look at it.

'Bake a chocolate cake, put some melted chocolate and laxative in the batter and mix it all up. Then get chocolate frosting and put the leftover chocolate laxative mixture in there, mix it up and frost the cake. Tell Elliot you made him a cake just to express how sorry you're about being a dickhead to him (that isn't the least bit true haha we all know that!) Sorry, bro...'
Dean Hensley

'For this prank you will need Febreze and a zip tie. You will need to tie the zip tie around the button of the Febreze. Then once Elliot is in the bathroom pull the zip tie and throw the Febreze in there and run. Have fun. Sorry, bro.'
Henry Free

'Next prank take Elliot to a public shower and then when he's in the shower run away with his clothes. Say u will meet him in the car so he will need to run to the car naked then lock the car door.'
Kristopher Stephen Peel

'Handcuff Elliot to his bed and cover his walls in pictures of himself to annoy him then superglue something to him.'
Bronwyn Peel

'Strap Elliot to a lamppost with a sign above him saying something entertaining or insulting to Elliot.'
Calum Sampson

SUGGESTION BOX

'You should superglue Elliot's feet to the floor.'
Esmeralda Bermudez

'VERY HOT – Know when he is likely to use the toilet? Rub some Deep Heat or any other type of warming muscle rub on the toilet seat!

STICKY – Unscrew the shower head and put some Kool-Aid Drink Mix powder behind the filter. Elliot will be surprised when a bright, sticky mess pours out.

COOKIES – Make some cookies with some finely chopped very hot chillies. Put the cookies out somewhere Elliot will find them.

HOT PEPPERS – Get some really hot peppers, Cut them up, smear pepper juice all over the door knobs in Elliot's room.
Good luck
#PoorElliot #SorryBro'
Jacob Corbin

'Okay I have two here, I don't know if Elliot is a big pop drinker, but if he is and uses ice cubes... make your own ice cubes, and put Mentos in them. Then make sure that Elliot takes those ones.
Now, you might have a door stopper on your wall to keep the door from scratching the wall. When Elliot's asleep one night, put an air horn on that door stopper thing, and when Elliot wakes up and opens the door, it should scare the hell out of him.
#Sorrybro #PoorELLIOT'
Jacob Corbin

'Hey I am a big fan, one idea is to connect a train horn to his phone and call him while he sleeps, another could be to fake send him a letter from the military.'
Joakim Lærum

'The other day my mum bought grass seed that had fertiliser in that looked like coffee granules or even chocolate chips. You could maybe prank Elliot by putting this in his coffee or even make a cake or something.'
Anthony Phillip

Say Cheese...
PARP

ELLIOT'S

We all know how my bro can kick off. When it comes to anger, Elliot has big issues. Knowing him as I do, I can tell when he's just a little bit upset and steaming mad. He only has to yell my name, in fact, and I can figure out from the volume whether I should carry on pranking or give him some space to calm down. I even made a chart so you can measure it for yourself.

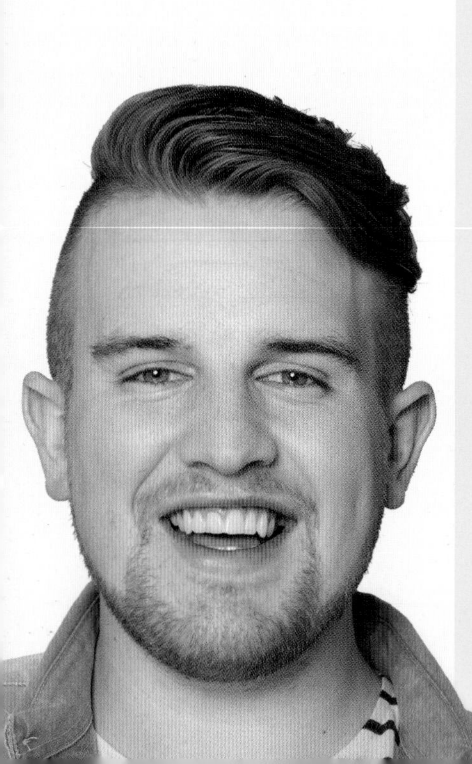

1 Silent Elliot

This is the moment that Elliot find out he's been pranked. From waking up in full make-up to discovering his bicycle sprayed hot pink, he'll stand there without word while he works out what's just happened. If you're very close to him at this stage, you can hear the cogs in his brain begin to turn. It won't end well.

2 Annoyed Elliot

This is marked by the first bellow. He'll call out my name sounding like a farm animal that's missed out on its feed. When I don't reply, it only gets worse. 'Ben . . . Ben . . . BEN . . . BEN!!!!'

ANGER-O-METER

3

Aggravated Elliot

Things could get physical from this moment. I've wound him up to the point where my bro is set to take matters into his own hands. Often this involves stomping to the shed so he can try to cut off stuff – like the rings I've stuck to his nipples – using gardening equipment. There's no stopping him now.

4

Furious Elliot

My bro is on the warpath, and I need to get out of his way. Here, Elliot has transformed from mild-mannered Welsh boy to a crazed lunatic. At the same time, his stupidity level starts to rise. He's a danger to himself.

5

Volcanic Elliot

Every now and then, my bro will hit this peak level of anger. The last time, when I shaved his head as he slept, he woke up and just went straight to this stage within the blink of an eye. Elliot started ranting at me so loudly that the police were called, armed with tranquilizer guns, because neighbours feared a bear had escaped from the zoo and hidden away in our house. My bro calmed down, thanks to the dart in his butt, and when he finally woke up he took one look in the mirror and blamed the barber. Result!

Chapter 9

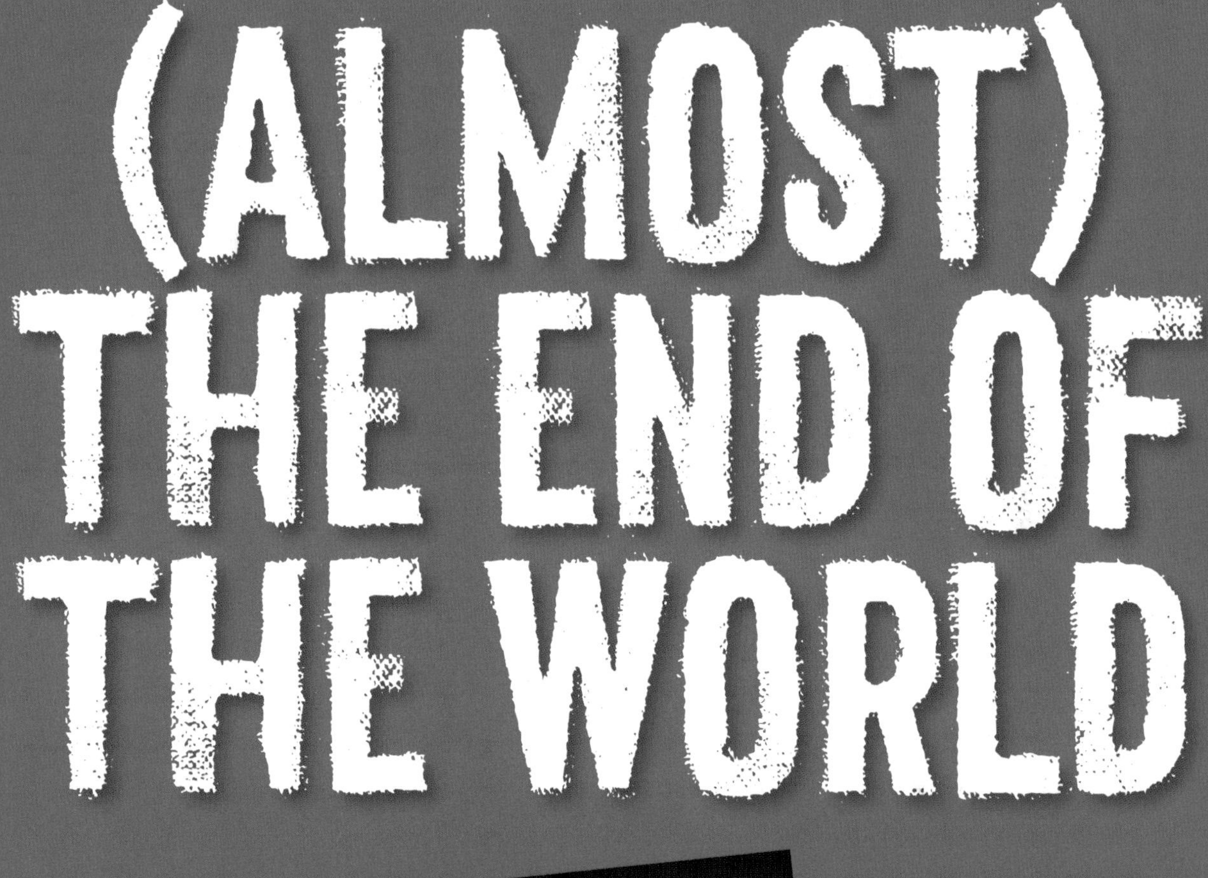

 **4.7 million–8.2 million**

I caught the whole drama on video. As planned, the rollercoaster guy had pulled the power switch with perfect timing. One moment the train was whooshing around the tracks, the next my brother and his replacement date were dangling from the top of the highest loop, and I couldn't stop laughing.

The sight of my brother in his clown wig, trapped and ranting at me, just made me crease up all the more. It was a special moment, and I had my social media powers to thank for cooking up a prank that would soon be shared around the world. I was struggling to get a grip when my man in the ticket booth quietly got the rollercoaster moving again and brought them back to earth. I zoomed in as Elliot raced to get away from Georgina's clutches.

At first I thought he was charging towards me. I couldn't help laughing, of course, but then he peeled away to one side. I swung my camera around to keep up, and found him face-to-face with Amelia. Her back was turned to me, but I could only think she must've found the whole thing hilarious.

Then, out of nowhere, she slapped his cheek.

'Oh,' I said to myself, and watched as she turned in tears and fled.

Without thinking, I stopped recording and lowered the phone. At the same time, Elliot spotted me through the crowd. I'd expected him to be so furious with me that his ears would be steaming.

Instead, he just looked utterly heartbroken. I'd set out to create the best clip ever, only to tread all over my brother's hopes and dreams.

'Ben, you went too far,' he told me the next day when I knocked on his door to sort things out. 'Amelia won't even talk to me now.'

'It was meant to be funny,' I said weakly. 'I didn't think she'd react so badly.'

Elliot stared hard at me. His eyes were a little puffy. It looked like he'd been crying, which just made me feel worse.

'She won't answer my calls,' he told me, 'and I can't really blame her. Who would want to go out with someone like me, knowing that a nutter like you was dead set on ruining EVERYTHING!'

It was a flash of anger, the first since Elliot's date crashed and burned, but it didn't last. How could it? Because straight away I promised him things would change.

'Bro,' I said, 'I've been thinking long and hard.'

'Did it hurt?' he asked.

'Elliot, listen to me,' I went on. 'What I did was meant to be a laugh, but it didn't end that way.' I paused there and took a deep breath. What I had to say next felt like it might hurt me, but not as much as the hurt I had caused.

Elliot looked at me in shock and surprise, and then his face seemed to brighten up.

'Really?' He paused for a moment. 'Do you mean that, or is this another joke?'

'I'm serious,' I said. 'I might have special social media powers, with millions of people looking to me to brighten their day, but this last prank has opened my eyes, Elliot. Why should you suffer so other people can feel better about themselves?'

It was a lot to take in, which is probably why my brother looked so confused for a moment. But then he nodded, and replied in a voice that was barely a whisper.

'Thank you, Ben.'

'It's the least I can do,' I told him, and then looked to the floor. 'Sorry, Bro.'

With my head bowed, I returned to my laptop and fired up my Facebook page. Overnight, even without the rollercoaster prank that I'd decided to delete, my follower count went past the eight million mark. As I prepared to write the most important post of my life, I sensed my social media powers try to stop me in my tracks. It was as if my fingers refused to move as I began to tap the keyboard, but I was determined to return to my old self.

Three hours later, with sweat beading my brow from the effort, I uploaded my farewell post.

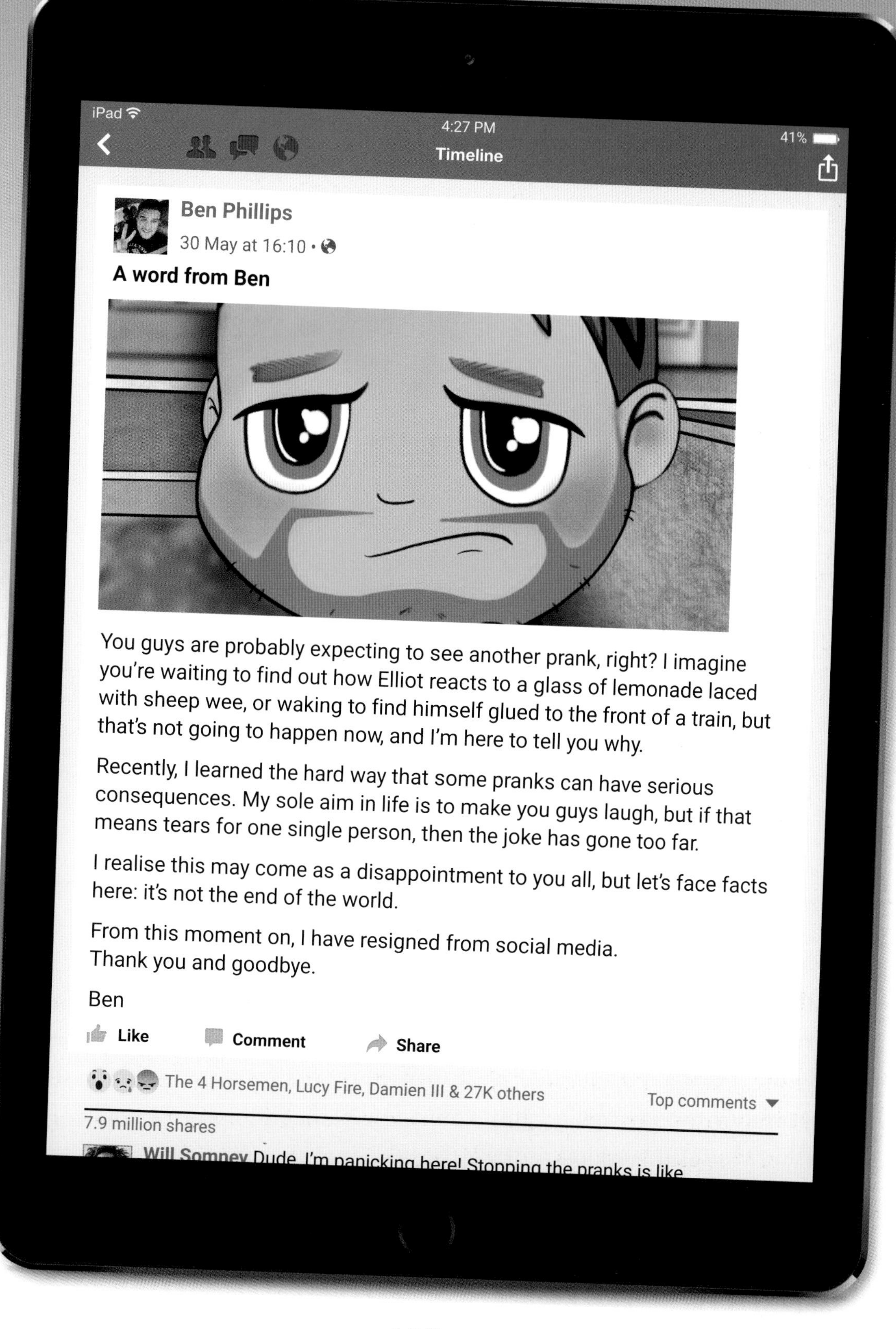
iPad
4:27 PM
41%
Timeline
Ben Phillips
30 May at 16:10 •
A word from Ben
You guys are probably expecting to see another prank, right? I imagine you're waiting to find out how Elliot reacts to a glass of lemonade laced with sheep wee, or waking to find himself glued to the front of a train, but that's not going to happen now, and I'm here to tell you why.
Recently, I learned the hard way that some pranks can have serious consequences. My sole aim in life is to make you guys laugh, but if that means tears for one single person, then the joke has gone too far.
I realise this may come as a disappointment to you all, but let's face facts here: it's not the end of the world.
From this moment on, I have resigned from social media.
Thank you and goodbye.
Ben
Like
Comment
Share
The 4 Horsemen, Lucy Fire, Damien III & 27K others
Top comments
7.9 million shares

The 4 Horsemen, Lucy Fire, Damien III & 27K others

Top comments

7.9 million shares

Will Somney Dude, I'm panicking here! Stopping the pranks is like switching off the sunshine. We need you guys!

Like • Reply • 7500 • 30 March at 15:45

879 Replies

Tallulah May As I read this, Ben, storm clouds have crept across the sky. Coincidence? I don't think so.

Like • Reply • 3200 • 30 March at 16:20

561 Replies

Shan Tine I used to make $65 an hour working from home. Now you've given up and my business has collapsed!

Like • Reply • 457 • 30 March at 16:23

244 Replies

Arianna Lilo Ever since you signed off, the temperature has plummeted here. It's like we're heading into a new Ice Age!

Like • Reply • 348 • 30 March at 16:27

180 Replies

Krash Dandy Reports of riots in the world's capital cities. Stay safe, everyone.

Like • Reply • 169 • 30 March at 16:29

64 Replies

Fred Wizz Are you guys for real? You spare your brother from further humiliation and civilisation starts to crumble. Save us, Ben!

Like • Reply • 42 • 30 March at 16:33

2 Replies

I didn't hang around to read the comments. My fingers ached from the effort of fighting my social media senses. Besides, there was one more thing I had to do. The phone I'd used to film every clip just felt all wrong in my hand. Instead of being a useful device, it seemed more like a reminder of a bad time I needed to forget. While Elliot ran a hot bath, safe in the knowledge that I wouldn't burst in and lob pink clothes dye into the water, I headed out across town to take care of my mobile in a fitting manner.

There was only one place for it, I thought to myself. The way I felt just then, the stupid phone was good for just one thing ...

I'D BEEN HERE ONCE BEFORE WITH GRANDAD.

IT WAS WHERE CLAPPE
OUT CARS CAME TO DI

ELLIOT'S DIARY

I'm still blimmin' cross with Ben. I'd found the girl of my dreams and he totally messed up my chances. Amelia doesn't just refuse to answer my calls or texts. The last time I tried to get in touch, it turned out she'd blocked me!

Still, it really does seem like Ben has learned his lesson. He can't stop apologising, in fact. All he ever says to me now is, 'Sorry, Bro!' like he means it, which is becoming quite annoying. In a way, I miss hearing him sound completely shallow about it, like he really doesn't care how I feel.

The fact is, ever since he closed his Facebook page and got rid of his phone, Ben has become a different brother altogether. And I'm not sure if I like it.

Sure, there are some plus points. I no longer have to look over my shoulder every minute, or wonder what surprise he has in store. It took a while to find the confidence, but now I can flush the loo without fear of pooey bubbles frothing across the floor because Ben filled the loo with washing-up liquid. Now, when I dive into a bowl of cereal, I can be sure the milk hasn't come from cousin Zoe's boobs (thanks, Ben), and it's been a while since I've come into contact with special glue. I'm even beginning to forget what it smells like (pigsties and regret, since you ask).

So yeah, life is good for me right now. It's just a shame I can't say the same thing for the rest of the planet.

The moment Ben pulled his Facebook page and left a fun-sized hole on the internet, things just took a turn for the worse. The sun has never shone that much over Bridgend, but now it's totally disappeared. The day after my brother put an end to the prank clips, dawn broke and then just gave up completely. We've been living in darkness ever since. Then the power cuts came, followed by the food shortages and the looting. When the TV channels stopped broadcasting, things got really serious. It left me with nothing to do but play a board game with my brother.

'It could be worse,' said Ben, as he rolled the dice and slid down the ladder for the second time in a row. 'I could be pranking you.'

I looked across at him, his face lit up by the flickering candle flame.

'You know, maybe the odd prank wouldn't hurt,' I offered. 'If it meant things went back to normal.'

'I have that power,' said Ben, 'but I'm not prepared to pie you to protect the future of the human race.'

I said nothing, taking my turn to roll the dice. It landed on a double six. I climbed the ladder all the way to the top.

'I win,' I said, but didn't sound convincing.

'Do you want to play again?'

'Thirty-six rounds of snakes and ladders is enough to last a lifetime,' I told him, and left the table to answer a knock at the door.

I'm not sure what I expected. Ever since the world began to go down the toilet, flushed away by my brother's decision to give up as the king of social media, I was ready for anything. Even monsters. If they were going to rampage over the hill and strike at my home town, now was the time.

'Hello?' I said through the gap in the door I had chained before opening. Peering through, I took one look at the growling, unsteady figure with the dead eyes and blood-soaked shirt, and decided on the spot that enough was enough.

'Ben!' I yelled, and closed the door. 'The zombie apocalypse has started!'

My brother appeared in the hallway. He looked a little scared. 'Already?'

'There's one at the door,' I said, and jabbed a thumb over my shoulder. 'To be fair, they probably feel at home in Bridgend.'

Ben looked around fearfully. 'What are we gonna do?'

My brother might've been more freaked out than me, but I was just unhappy. I didn't want the world as we knew it to come to an end. I wanted to watch TV and smell fresh flowers in the park. OK, I wasn't bothered about the flowers, but you know what I mean. More importantly, with the supermarket shelves standing empty, I was worried I might go into some kind of critical biscuit withdrawal.

While Ben looked trapped and a bit defeated, I knew we still had one option. So I waited until I had his full attention before I told him what I had in mind. 'I'm going out for a bit,' I told him, before making my way to the back door. 'I might be some time.'

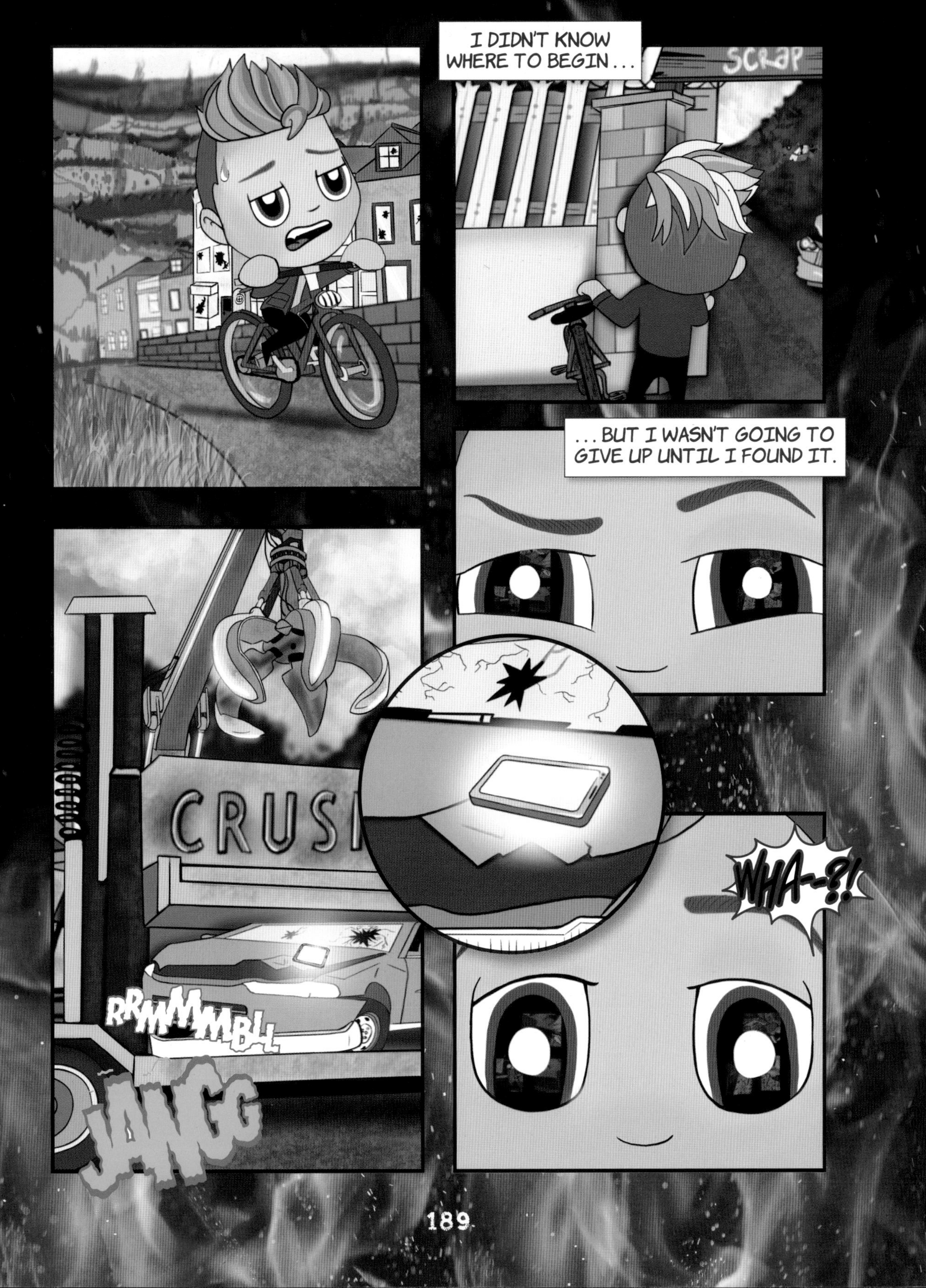
I DIDN'T KNOW WHERE TO BEGIN . . .
SCRAP
. . . BUT I WASN'T GOING TO GIVE UP UNTIL I FOUND IT.
CRUSH
RRMMMMBLL
JANGG
WHA--?!

I HAD JUST SECONDS TO SAVE THE WORLD . . .
. . . AND I DID IT!
BA-DOOM
RRMMMBLL
NO! MY BIKE!

I got back home a little later than planned. For a start I had to walk, which meant dodging zombies. Ben was waiting for me. He looked worried sick.

'Where have you been?' he asked.

'Never mind.' Without further words, I handed him his phone. For a moment, he held it in the palm of his hand as if unable to take in what I'd done here. We also had a signal. Just a bar that flickered on and off, but it would be enough. We were used to it, to be honest. This was Bridgend, after all. The land that 4G forgot. Finally he looked back up at me. 'You know what you have to do,' I told him.

'Call the police?'

'No,' I said and thumped him on the shoulder. 'The future is in your hands, Ben. Only you have the power to unite everyone by making them laugh. It's the surest way to lift their spirits and bring light back into the world.'

'But Elliot,' he said, only for his voice to trail away.

I stared at him and began to nod. He knew what I was thinking. It was the only chance we had to save the human race from misery and extinction.

The Moaner Geezer

Benatello

MAKE YOUR OWN PRANK CHANNEL

1 CHOOSE YOUR PLATFORM

I'm a Facebook guy, but you can always post some things on other platforms, and then share the content to reach as many people as possible. It really is down to you. This is your show!

2 YOUR PROFILE PICTURE

Don't look so miserable! This isn't a school photo. It's your shop front, and you're inviting people in to look around. So, c'mon – smile!

Don't leave it to me to have all the fun. Find yourself a target that's not going to burst into tears or take you to court, and get up to the kind of tricks that could earn you online fame.

~~SILLY PRANKERS~~
~~THE BROTHERS~~
BEN PHILLIPS

3 YOUR CHANNEL NAME

Let's face it, 'Elliot's Special Channel' sounds all wrong on so many levels, which is why my bro's page withered and died of neglect. Be creative. Show some spirit!

4 CREATE YOUR CONTENT

This is where the fun begins! A smartphone is all you need to start recording clips to upload. Then you just need to come up with some cracking pranks that people will talk about and share.

WHAT'S IN

They say that the stuff you carry with you at all times reveals what you're really like. I'm happy to show you mine, then let's wait for Elliot to fall asleep on the sofa and take a peek at what he's packing.

ELLIOT

Sticking plasters

Nipple-soothing gel

Make-up removal wipes

Mankini (pink . . .)

Deely-boppers

Fluffy handcuffs

Wallet including photograph of a badger (what?)

Accident & Emergency admission form

Fish and chip shop loyalty card

Creased picture of Katherine Jenkins torn out of a magazine (hmm . . .)

YOUR BAG?

BEN

Whoopee cushions
Hot sauce
Fake moustaches
Special glue
Plastic vomit
Nipple clamps
Duct tape
Car keys
Georgina's dress (eh?)

Basket Case

Elliot's always dashing to the supermarket to pick up plasters, ointment or nipple-soothing cream. But once he's there my bro quickly turns his attention from first aid to food. As someone who thinks 'five a day' is all about naps, he has a lot to learn. So the last time he left me to guard his bag while he nipped to the loo, I decided to look through the contents.

And what I found left me wondering what goes on inside Elliot's mind.

Crisps
Nuts

BRIDGEND 24/7 STORES

Mags, Fags, Bags, Rags, Tags...
(Closed Wednesdays)

1 apple	0.65
2 bagels	1.69
5 packets of jammy dodgers	3.75
1 packet of cornflakes	1.29
1 pint of orange juice	1.09
1 bag of donuts	1.99
Pack of pork chops	4.70
1 waterproof bedsheet	12.99
1 nasal hair trimmer	17.99
1 pack of disposable rubber gloves	3.99
4 bottles of bleach	5.16
22 packets of Prawn Cocktail crisps	10.78
2 packs of TastiNuts Vegan Munch Bars	3.58
Cat repellent spray	2.99
One toilet plunger	1.89
TOTAL	74.53

Half eaten – no surprise.

Elliot always asks for a discount because of the holes.

2 packs opened – he's an animal!

Opened – Elliot! Can't you wait until you get home?

Opened – the healthy option.

Opened – the not-so-healthy option.

Opened – Bro, you're supposed to cook raw meat!

Hello?

Eeyuuuw!

What?

Has he murdered someone and plans to dispose of the body?

I hate that flavour.

Must've fallen into his basket by accident.

You really hate Norman, don't you?

Which you can expect to find stuck to your forehead when you wake up in the morning. Sorry, Bro!

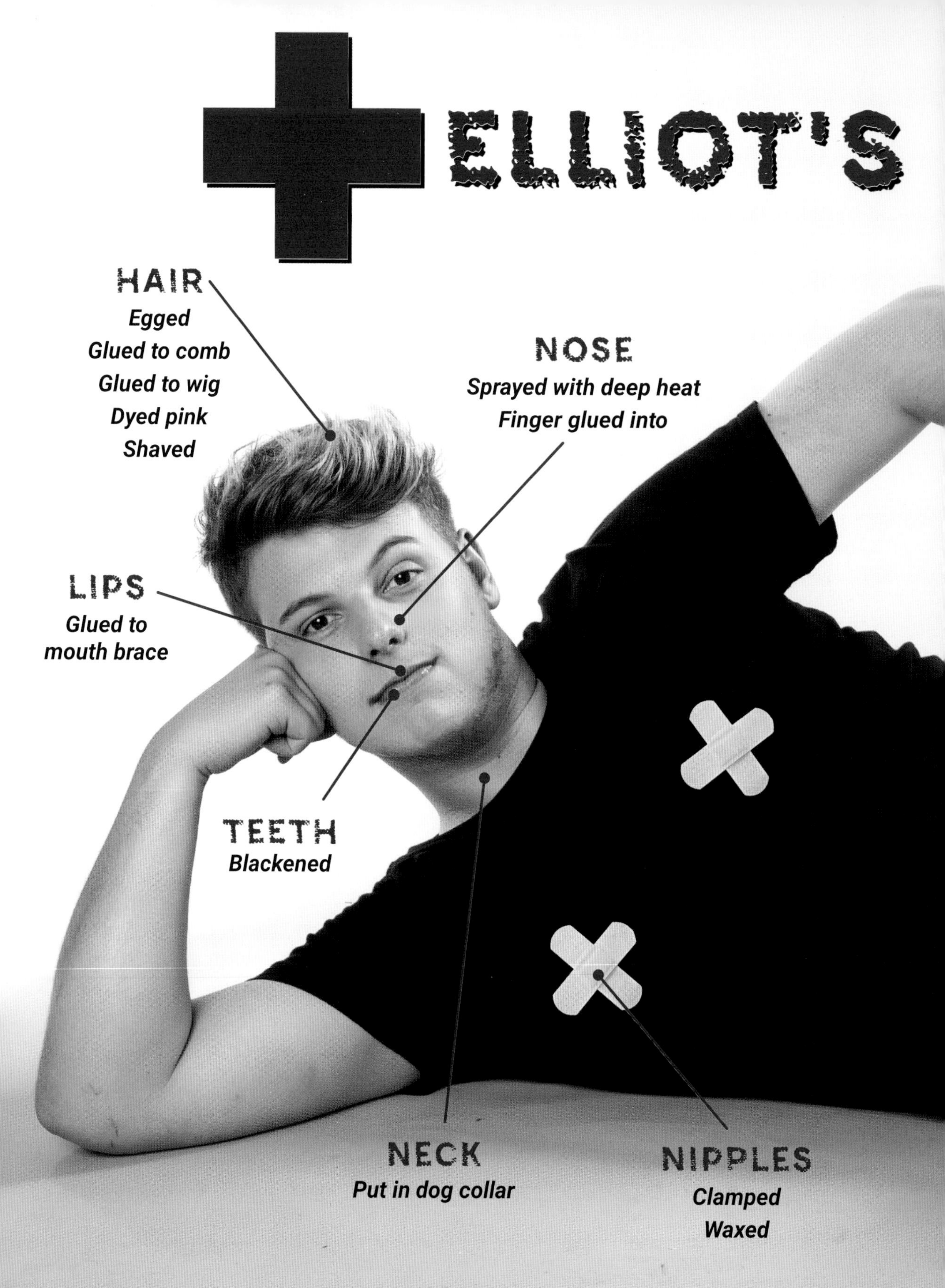
ELLIOT'S
HAIR
Egged
Glued to comb
Glued to wig
Dyed pink
Shaved
NOSE
Sprayed with deep heat
Finger glued into
LIPS
Glued to
mouth brace
TEETH
Blackened
NECK
Put in dog collar
NIPPLES
Clamped
Waxed

INJURIES

SKIN
Moisturised with honey and bodily fluids
Dyed blue
Dyed with industrial tanning lotion
Showered in chilli
Bleached with bathroom wipes
Pierced

FINGERS
Broken

BUM
Stuck to toilet seat
Rubbed in wasabi paste

FEET
Stuck to flip-flops
Stuck to high heels

PANTS
Washed with stinging nettles
Rubbed in itchy house insulation

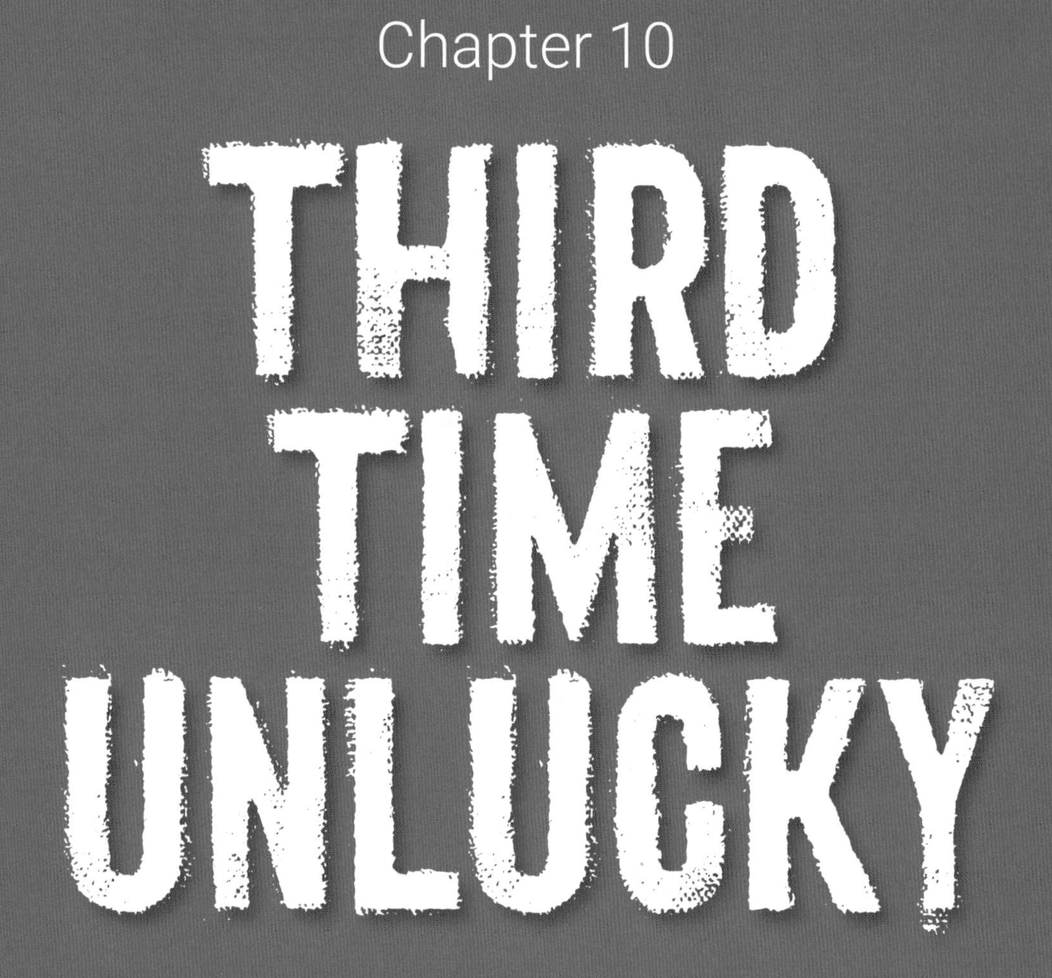

Chapter 10

THIRD TIME UNLUCKY

8.2 million–10 million+!!

With my Facebook page back in action, and a new clip uploaded, the sun came out from behind the storm clouds. Those first bars of light were enough to drive the zombie hoards over the hills and far away, while birdsong filled the air along with the sound of laughter. It was so nice to hear people chuckling. The birds were just as annoying, but still.

With power restored across the world, people from every nation united to laugh at the clip of Elliot falling for a dare to climb onto a bull belonging to Farmer Chris, only to discover that I had coated the saddle in special glue.

It goes without saying that Elliot was unhappy, but not half as much as the bull. That angry beast made every attempt to buck and toss him into the mud. When he finally came loose, and landed on his face in a pile of manure, I just knew we had the perfect prank to save mankind, and that was enough to calm my brother down.

'You're a star,' I reminded him a few days later as my follower count returned and climbed towards ten million.

'My bum is so sore I can't sit properly,' he grumbled, 'but I suppose it was worth it.'

I'd made Elliot a cup of tea. He was sitting on a rubber ring in the front room. With a black-and-blue backside, it was more comfortable than the sofa, apparently. For once, I didn't mock him mercilessly.

'I want to thank you, bro,' I said, and sat across from him. 'You're like a modern-day saint. I wouldn't be surprised if they put up a statue of you in town.'

Elliot smiled weakly.

'No doubt you'd deface it within days,' he said.

'Well, yeah,' I admitted. 'Maybe some glasses and a moustache, but I mean it when I say that if there's anything I can do to make it up to you, I'll make sure it happens.'

'Anything?'

Elliot considered me carefully.

'Within reason,' I added.

'There is one thing,' he said after a moment, only to wave away whatever he was thinking. 'Forget it,' he added. 'That'll never happen.'

No lie, I read his mind just then. I knew exactly what Elliot was hoping for more than anything. Just then,

I swore to myself that I'd make it happen. It took a great deal of effort, mostly in the form of some very humble grovelling to a girl I'd upset so badly. I even promised Amelia that if she got back together with my brother she wouldn't come out of this tarred, feathered, covered in paint, slime or slurry. I simply wanted to make up for everything, and give love a chance to shine.

I know. Smooth, right? They don't call me Flirty Phillips for nothing. In fact, they don't call me that at all. But it worked a treat. As I explained that Elliot had been the boy who saved the world from oblivion, her frosty expression melted and she even began to swoon. It was a surprise to find that Amelia could be this old-school when it came to romance. I also had no clue what on earth she saw in Elliot, but I was determined to help bring them together.

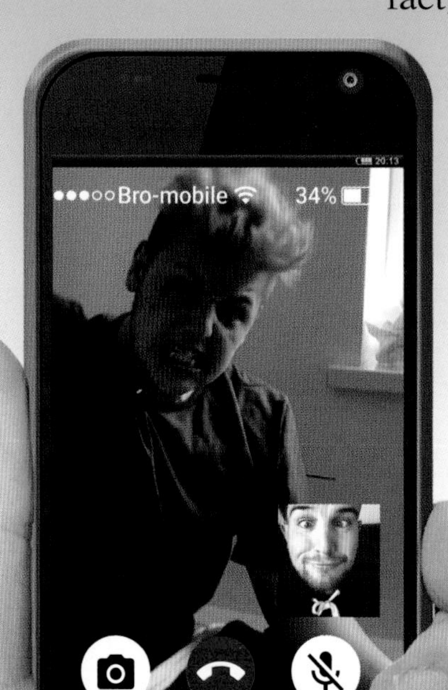

'I'm gonna write to him,' she declared, and pressed the back of her hand to her brow.

'Can't you just text?' I suggested. 'You could also try FaceTime, but be careful with that in case you call and he's on the loo. Elliot forgets himself sometimes, and you really don't want to be looking at him with his straining poo face on.'

'Ben,' said Amelia without even blinking. 'I'm doing this properly.'

So I just waited quietly while she penned him a letter with an invitation he couldn't refuse. I didn't like to say there was no need to be this soppy. I even told her she could trust me to give it to him personally once she'd sealed the envelope. Naturally, I opened it on the way home. What I read told me this was a date that could only help Elliot to forgive me for a lifetime of pranks and humiliation.

'Bro,' I said, on handing him the envelope. 'You're going to love this.'

Hand on heart, I wanted this to be a truly special event for Elliot (while entertaining the world as I quietly live-streamed every moment across the internet).

Rude not to, right?

To my bae . . . Elliot,

OK, I'm gonna start by getting this off my chest. What happened at the fairground broke my heart! Who was that girl with the beard? In fact, don't answer that! Just promise me you'll have nothing to do with her ever again – that's if you want a second chance with me.

Basically, Elliot, your brother has explained everything to me. I realise now you didn't plan any of it, and though it looked like you and that girl were getting frisky with each other, I'm prepared to believe Ben (for once!) He also tells me that you proved yourself to be a bit of hero recently! And I do so love a hero, Elliot. Especially one who makes the whole world smile again.

So how about we work things out together over dinner? Ben tells me that you love nothing more than fish and chips from your local takeaway, which sounds good enough to me. He's even promised to make sure they give the place a good clean in our honour.

So what do you reckon, babes? I've missed you!

Hugs,

Amelia

ELLIOT'S DIARY

Ben really thinks I'm stupid. I finished reading the letter he gave me and balled it into the bin.

'I'm not falling for it,' I told him. 'Faking a letter from the love of my life isn't funny. You're a sicko!'

'I swear it's from her!' Ben picked the letter from the bin and flattened it out. 'Look at the handwriting, Elliot. It's all girly-swirly, and there's no spelling mistakes. It even smells of blimmin' perfume! C'mon, bro. Do you really think I'm capable of that?'

I took another look. At the same time, I couldn't ignore the scent that came off the page. I looked up at my brother. He didn't even smirk.

'Is it really on?' I asked.

'Bro', he beamed and spread his hands wide. 'Prepare to score like a Welsh striker in front of an open goal!'

'That's no guarantee,' I pointed out.

'You know what I mean.'

To be fair to Ben, he went out of his way to help me get ready. He took me down to the suit hire shop and sorted me with an outfit that he said Amelia would find impossible to resist.

'I can barely breathe,' I complained as he did up my tie. 'Are you sure this is her thing?'

'Totally,' he said, before persuading me to nip to the barber's for a trim. 'Just wear a belt for once in your life so your boxers don't hang out the back.'

'But it's comfortable.'

'It's wrong, Elliot. That's what it is.'

Later that evening, when my brother led me to the chippy, I went back to thinking this was a massive wind-up.

'I like a Friday Special with extra vinegar,' I told him, 'but this is hardly the most romantic venue.'

Ben silenced me by swinging open the door. Normally, the chippy had the radio blaring. So it was a surprise to see the pianist in the corner over by the ketchup and serviettes. As soon as I stepped inside, he began to tinkle away. It took a moment for me to take in the candlelight and the rose petals scattered across the counter. In shock, I looked at Ben. He was just flipping the OPEN sign to CLOSED.

'It's just for you,' he said, and then gestured towards a corner table. 'And your date for the evening.'

'Amelia!' There she sat, looking like an angel with greasy elbows. At least that's what usually happened if you leaned on the tables in here, but tonight, it seemed, the place has been properly scrubbed.

'It's so good to see you.' She rose from her chair to kiss me on both cheeks. I blushed and tucked in my shirt a bit tighter. 'This evening is going to be magical, Elliot, and it's all thanks to Ben.'

'No need to thank me, bro,' he said, backing away as if to give us space. 'Who'd have thought a simple lad could save the world, but you did it, Elliot, and this evening is all about you!'

I frowned as he retreated behind the counter and even bowed as he went.

'What do you mean by "simple"?' I asked, but by then he had disappeared.

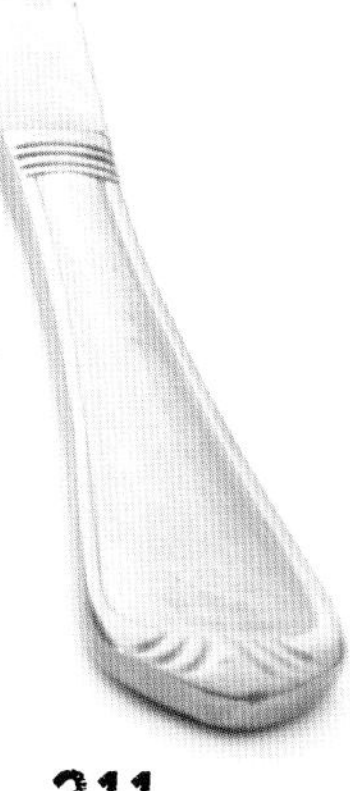

HAVE YOU DECIDED WHAT YOU'D LIKE, ELLIOT?
DR HARLEY STREET
Battered in Bridgend
I'LL HAVE THE FISH . . .
. . . AND THE CHIPS, I THINK. HOW ABOUT YOU?
WHATEVER ELLIOT IS HAVING SOUNDS DELICIOUS TO ME.

HERE'S TO US, BABES!
AND HERE'S TO BEN FOR BRINGING US TOGETHER AT LAST.
WHAT ELLIOT DOESN'T KNOW IS THAT I CAN'T POSSIBLY WALK AWAY FROM THIS PRANK OPPORTUNITY!
Instant Laxative
LET'S SEE HOW ELLIOT GETS ON . . .
Instant Laxative
PLIP PLIP
SPLOOSH
The End
OH!
I SHOULDN'T ADMIT IT BUT . . . YEAH.
WHAT'S WRONG? ELLIOT, HAVE YOU JUST POOPED YOURSELF?
EUUUWWWW!!!
SNIGGER!
BEN? BEN . . . WHY YOU RECORDING? OI! BEN!!!
SORRY, BRO!!

Ben Phillips
@BenPhillipsUK
It's not all pranks, I love him really 😈😡 #sorrybro
8 Sept 2016

Elliot Giles
@ElliotGilesUK

One day I SHALL get my revenge 😬😬😬

8 Sept 2016

AN IDIOT'S GUIDE TO PRANK FILMING

When is a prank not a prank? The answer is when your target spots the camera just moments before they bite into that chilli-laced hotdog or look in the mirror and see the measles spots you've drawn all over their face. If they work out what's going on then the joke is on you. That's why it's so important that you keep your camera undercover and out of sight. Here's how to set up everything so you don't look like a fool.

Do your homework

Yes, I know. This isn't the kind of advice you'd expect to hear from me, but when it comes to pranking it's well worth putting in some advance work. You need to suss out the scene, whether it's the bedroom, the kitchen table or your local park. Think about every angle, and work out the best place to capture the prank on film.

Ben's tip: take your time and be creative.

Plot the action

The scene of your prank is a bit like a film set. The only problem is your star has no idea they're being filmed. It means you don't have much control over where they'll stand or how they'll react. So think of the best place to record the moment it all goes wrong for your victim, so that you capture the action and the reaction.

Ben's tip: set up your camera on wide focus. That way, it'll record everything.

Be camera-ready

Once you've worked out a good hiding place for your camera, and you know that nothing's going to get in the way, take steps to secure it in place. You don't want to risk the camera being knocked over so it films the sky. Nobody wins views with that kind of thing. Apart from cloud-watching channels.

Ben's tip: pack string, crocodile clips or rubber bands. Anything to keep that camera nice and steady.

Why you recording?

Your camera's in place, the scene is set and your target is just seconds away from being pranked. If this is the moment you realise that you haven't pressed the record button then it's time to have a long, hard think about where you just went wrong. It's an easy thing to forget, but could possibly cost you millions of views.

Ben's tip: make sure you have enough storage space to record, and hit that button in good time.

ELLIOT'S GUIDE *to* PRANK AVOIDANCE

At last! It's my turn to tell you that pranking is NOT FUNNY! Who wants to be pied in the face, put on a pair of pants full of ants or find TEAM GOAT tattooed across their chest? Not me, that's for sure, but it's never stopped Ben from making my life a misery.

So if there's a Ben in your life, this is my chance to help protect you from their pranks. Follow my handy guide to staying safe and you'll never have to visit A&E with a ginger wig stuck to your privates.

Always sleep with one eye open

OK, so I'm not very good at this. But I'm practising! Ben always pranks me when I'm dozing.

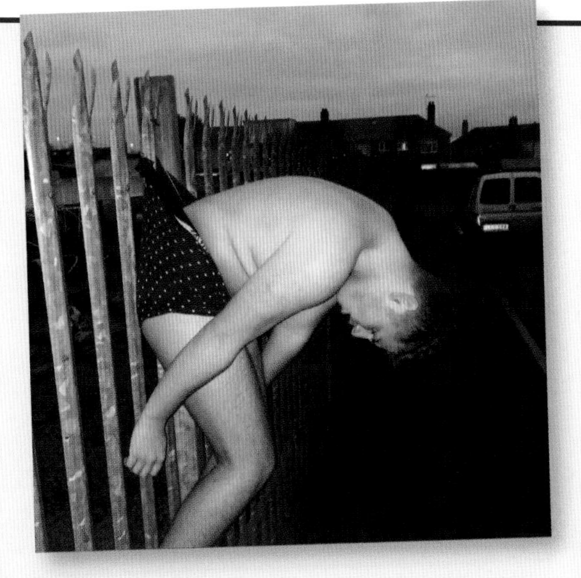

That's why I've started sleeping with a matchstick jammed between the lids of one eye. I'm smart, see? And half blind now as well, but hopefully it's only temporary.

Trust nobody

I get a lot of hot girls inviting me on dates. It's true! Sadly, most of those texts come from my brother. He's a master at setting up fake dating profiles, but I'm not falling for it any more. I've decided I can't trust anyone. Recently, when Nana and Grandad tried to give me ten quid to spend on something nice, I turned them down. Why? Well, I dunno. I just figured it was all part of some stupid prank that my brother might try to pull on me. Now who's laughing, eh?

Make them taste your tea before you eat it

On TV recently I learned that Roman leaders used to hire people to sample their feasts in case someone had poisoned them. I'm not saying Ben wants to kill me, but if my meal looks fishy I make him try a mouthful.

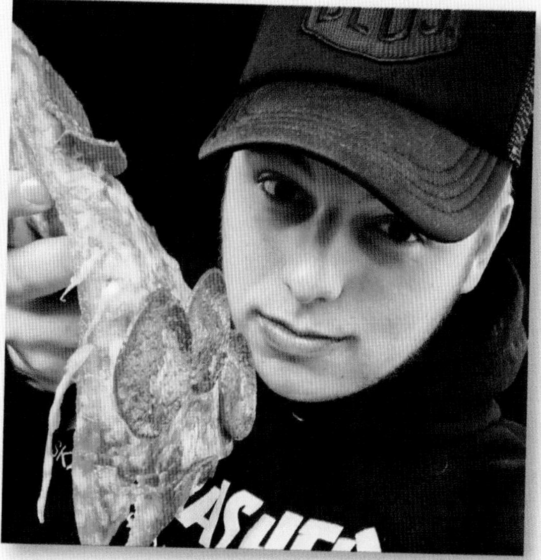

Don't get mad. Get really, really mad!

Next time you get pranked, don't go whining to a grown-up. Sort it out yourself! If you're anything like me, there's a monster inside just waiting to break out. And if there's one thing that scares Ben, it's my angry face. He might be laughing on the outside, but deep down I know he's bricking it. One day he's gonna push me too far. Then he'll be sorry. So don't make me angry, Ben. Don't push my buttons or wind me up the wrong way. Just leave me alone and STOP RECORDING!!

SUPERHERO! SUPERZERO!

TO THE PRATMOBILE!
CURSES!

MY SIDE KICK, HA HAA!

FWAP
KRAK
ARRRGGH!

PLAN YOUR OWN PRANK

Great pranking takes planning, so make notes below on what you want to do and draw your blueprint around Elliot.

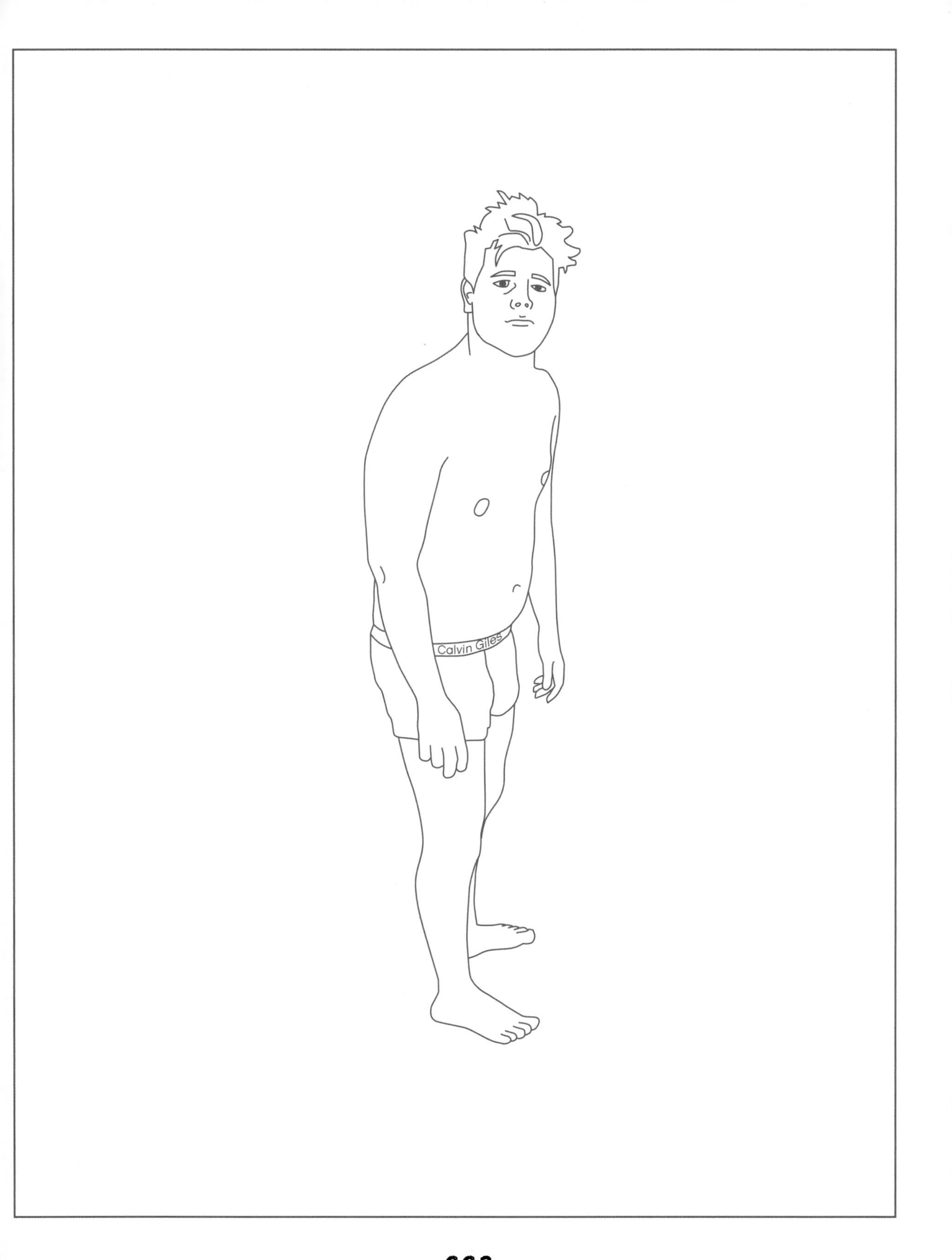
Calvin Giles

THANK-YOUS AND SORRYS

Sorry Bro, for writing this book without your permission . . . but **thanks** for being the best friend I never had.

Thanks Mum, for finding Elliot in a bush and making this book possible.

Sorry Nan, for Elliot being a disgrace.

Thanks Dad, for meeting Mum and making me.

Sorry Farmer Chris, for introducing you to Elliot (but **thanks** for letting me borrow your tractor).

Thanks Georgina, for making the pranks epic.

Sorry too, future Ben Phillips, for any potential cringe moments when you look back on this.

Thanks Grandad, for being hilarious!

Sorry cousin Zoe and Simone, for Elliot harassing you.

Thanks Gay Uncle Vic, for supporting Elliot.

Sorry Dodgy Uncle Ian, for having to be our criminal adviser.

Thanks Nutty Uncle Anthony, for the biscuits and tea every Saturday.

Sorry Wise Uncle Rob, for you having to be a member of our family.

Thanks Playboy Aunty Liz, for getting us a free subscription.

Sorry Foot Doctor Aunty Sarah, for having to cut Elliot's smelly toenails.

Thanks Druggy Uncle Barry, for keeping us on a high.

Sorry Jordan.

Thanks Grumpy Grampy, for teaching me how to fish.

Sorry not **sorry** Joe, for introducing me to the grapevine.

Thanks Rachael Bevan, for making me look sexy in the drawings, and Elliot . . . well.

Sorry Adam Lynch, that you have to hold mine and Elliot's hands 24/7.

Thanks Natasha Hambi, for being our London mother.

Sorry Alex Bewley, Gordon Hagan, Dave Levy . . . you're stuck with me for a long time.

Thank you to WME and Facebook.

Most importantly, **thanks** to every single one of you who followed me from day one, when I was making Vines by myself and with the little man, all the way to recording poor Elliot today. I don't need a morning alarm, a quick fix, or an energy shot . . . I just look at my notifications and see the love you send. It keeps the butterfly in my stomach floating!

Why am I recording? It's simple: you've not only changed my life, but millions of others have benefited from the videos bringing them happiness in bad times. Every time you liked and shared, you passed the message on, and you brought someone joy . . . And you helped me personally out of my darkness when I felt my life was at a standstill. Thousands of you gathered together, and shouted loud, and kept spreading the word. **Thanks** to the fan pages that have made my viewers into a huge family – you really connect the dots. Even if you screenshot my ugly snapchats . . .

Remember: no matter what blocks your path or dims your light, it could be worse: you could be Elliot. But I couldn't imagine going on this journey with anyone except him . . .

Thanks Bro, for believing in me when nobody else did, and being there when I was just poor Ben. No matter what I've put you through, you've been the first one to put your arm around me and say we're gonna be OK.

KEEP SMILING!

Ben Phillips